Jeffrey H. Miller

DEPARTMENT OF MICROBIOLOGY
AND MOLECULAR GENETICS
AND MOLECULAR BIOLOGY INSTITUTE
UNIVERSITY OF CALIFORNIA LOS ANGELES

Discovering Molecular Genetics

SOLUTIONS MANUAL & WORKBOOK

Cold Spring Harbor Laboratory Press 1996

Discovering Molecular Genetics

SOLUTIONS MANUAL
& WORKBOOK

Design by Emily Harste

Front cover: Original black and white drawing by Margaret L. Kowalczyk based on an idea suggested to the author by Anthony J.F. Griffiths.

ISBN 0-87969-482-3

10 9 8 7 6 5 4

All Cold Spring Harbor Laboratory Press publications may be ordered directly from Cold Spring Harbor Laboratory Press, 500 Sunnyside Boulevard, Woodbury, New York 11797-2924. Phone: 1-800-843-4388 in Continental U.S. and Canada. All other locations: (516) 422-4100. FAX: (516) 422-4097. E-mail: cshpress@edu. For a complete catalog of all Cold Spring Harbor Laboratory Press publications, visit our World Wide Web site http://www.cshlpress.com/

Contents

Preface

The following pages contain the solutions that I have worked out for the problems and scenarios in Section III of *Discovering Molecular Genetics*. There are certainly alternative ways to solve some of these problems, but the solutions given here are my recommended answers. Some of the Poisson distribution scenarios ask for the probability that a certain number appears, for example, 21 times among many spins of the roulette wheel, when the average is, say, 10. In a strict sense, we would really want to calculate the probability of a number coming up 21 times or more, so we would sum the probability for 21, 22, 23, etc., until the increment becomes so small that it is negligible. Usually that means calculating only one or two additional terms. However, in these problems we have not done that, but instead have simply calculated the probability for the one number, in this case 21.

We have included as an appendix some of the actual student answers to several of the scenarios, to show the type of work that students taking this course at UCLA and other universities can generate in response to some of the scenarios.

This solutions manual would have been impossible to prepare without the tireless efforts of Nancy Ford at Cold Spring Harbor Laboratory Press. However, I am responsible for any errors in the solutions, and I apologize if any of these slipped through.

May 14, 1996

Jeffrey H. Miller
University of California
Los Angeles

Solutions to Unit Problems

Unit 1 / The Fine Structure of the Gene

1. Here, we are given the fact that 400 of 1000 tubes had no β-galactosidase molecules and are asked to calculate how many tubes had 1, 2, 3, and 4 molecules, respectively. We can calculate:

   $P_0 = 400/1000 = 0.4 = e^{-m}$,

   and thus,

   $\ln(-m) = \ln(0.4) = -0.91629$, or $m = \mathbf{0.916}$.

   Therefore,

   $P_1 = me^{-m} = 0.916 \times 0.4 = 0.3664$,

   which is the **fraction** of tubes with 1 molecule of β-galactosidase. To get the expected **number** of tubes with 1 molecule, we multiply this fraction by the total number of tubes, which is 1000, to get **366** tubes with 1 molecule each. Similarly,

   $P_2 = m^2e^{-m}/2 = (0.916)^2 \times 0.4/2 = 0.1678$,

   and multiplying by 1000 gives **168** tubes with 2 molecules each;

   $P_3 = m^3e^{-m}/6 = (0.916)^3 \times 0.4/6 = 0.0512$,

   and multiplying by 1000 gives **51** tubes with 3 molecules each;

   $P_4 = m^4e^{-m}/24 = (0.916)^4 \times 0.4/24 = 0.0117$

   and multiplying by 1000 gives **12** tubes with 4 molecules each.

2. **(A)** In this problem, because we think there are hotspots, we cannot count all 100 sites as being part of the distribution. Thus, we must avoid the trap of determining m by dividing the 400 mutations by the 100 nonsense sites. Moreover, we are given the number of sites with 5 and 6 occurrences and told to consider them as part of the normal category of sites.

   a. Given the above, we use the two adjacent points with 5 and 6 occurrences to calculate m as follows:

   $$P_6/P_5 = m/6 = \frac{\text{no. of sites with 6 occurrences}}{\text{no. of sites with 5 occurrences}} = \frac{15}{15} = 1.$$

   Thus, $m/6 = 1$, and $m = \mathbf{6}$.

   b. To find the number of sites with 0 occurrence, we can use the ratio $P_6/P_0 = m^6/6! = 6^6/6! = 64.8$, which also equals the ratio of the sites with 6 occurrences to those with 0 occurrence. Rearranging gives the number of sites with 0 occurrence as $15/64.8 = \mathbf{0.23}$.

   At this point, we should calculate the total number of sites in the random distribution. Since $P_0 \times$ the total = 0.23, the total = $0.23/e^{-m}$. With $m = 6$, this gives **92.8** (which we would round off to 93) as the total number of sites.

   c. Having calculated the total number of sites, we can then calculate that the expected number of sites with 9 occurrences = $P_9 \times$ the total. For $m = 6$, $P_9 = 0.0688$, which multiplied by 92.8 = **6.4** as the expected number of sites with 9 occurrences.

Likewise, the expected number of sites with 14 occurrences = $P_{14} \times 92.8 = .0022 \times 92.8 = \mathbf{0.2}$. Therefore, in principle, there is an 80% chance that the site with 14 occurrences is a hotspot.

(B) As in Problem 2A, we use the two adjacent points with 4 and 5 occurrences to calculate *m*.

a. $P_5/P_4 = m/5 = 14/14 = 1$. Thus, $m = 5$.

b. The expected number of sites with 0 occurrence is **0.538**, and the total number of sites is 79.8 (rounded off to 80).

c. The expected number of sites with 10 occurrences = 0.018 x 79.8 = **1.44**.

The expected number of sites with 12 occurrences is **0.274**, and the chance of a site with 12 occurrences being a hotspot is about 73%.

(C) As in Problem 2B, we use the two adjacent points with 4 and 5 occurrences to calculate *m*.

a. $P_5/P_4 = m/5 = 16/16 = 1$. Thus, $m = 5$.

b. The expected number of sites with 0 occurrence is **0.614**, and the total number of sites is 91.2 (rounded off to 91).

c. The expected number of sites with 11 occurrences = 0.0082 x 91.2 = **0.75**.

The expected number of sites with 13 occurrences is **0.12**. The chance of a site with 13 occurrences being a hotspot is about 88%.

3. Here, we have a gene with 1200 base pairs and 15 base-pair intervals. The number of intervals is thus 1200/15 = 80.

(A) There are 225 deletions and we can calculate the expectation for a random distribution with these deletions being distributed into the 80 intervals. In this case, $m = 225/80 = 2.8$.

Since the expected number of intervals with 14 deletions is P_{14} x the total = $1.27 \times 10^{-6} \times 80$ = **0.0001**, which is less than 0.001, we do consider the interval to be a hotspot for deletion endpoints.

(B) Here, there are 200 deletions, and in this case, $m = 200/80 = 2.5$.

Since the expected number of intervals with 12 deletions is P_{12} x the total = $1 \times 10^{-5} \times 80$ = **0.0008**, which is less than 0.001, we do consider the interval to be a hotspot for deletion endpoints.

(C) Here, there are 300 deletions, and in this case, $m = 300/80 = 3.75$.

Since the expected number of intervals with 15 deletions is P_{15} x the total = $7.33 \times 10^{-6} \times 80$ = **0.00059**, which is less than 0.001, we do consider the interval to be a hotspot for deletion endpoints.

4. **(A)** Because we know that some of the 350 sites in the gene are hotspots, we cannot calculate *m* by counting all of these sites as part of the total, particularly since we are given the information that we are considering the sites with 2 and 3 occurrences as part of the distribution involving sites with equal frequencies of mutations. There are 60 and 20 sites, respectively, with 2 and 3 occurrences. To determine whether any of the other sites are hotspots relative to these sites, we calculate the distribution based on the value for *m* that we get by using the occurrences at these sites.

a. $P_3/P_2 = m/3 = 20/60 = 1/3$. Thus, $m = 1$.

b. We can get the expected number of sites with 1 occurrence by the formula $P_2/P_1 = m/2$ = number of sites with 2 occurrences/number of sites with 1 occurrence. Thus, 60/sites with 1 occurrence = 1/2. Rearranging gives the number of sites with 1 occurrence = **120**.

c. To determine how many sites are **not** hotspots, we simply determine the total number of sites in this distribution. Since P_1 x the total = 120, the total = $120/P_1$, and since $P_1 = 1 \times e^{-1}$ = 0.37, the **total = 324**.

(B) Considering what is given, we can calculate the expected number of genes with any number of insertions if everything were random. For this random distribution, $m = 350/500 = 0.7$, since the insertions are being distributed among the genes.

a. From the chart provided in the text for Problem 4A, for $m = 0.7$, $e^{-m} = 0.5$, and the number of genes with 0 insertion would be P_0 x the total, or $e^{-m} \times 500$ = **250**.

We can calculate each of the next points using the ratio P_1/P_0 = number of genes with 1 insertion/number of genes with 0 insertion = $m = 0.7$. Thus, the expected number of genes
with 1 insertion = **175**
with 2 insertions = **61.25**
with 3 insertions = **14.3**
with 4 insertions = **2.5**.

b. Since we expect to find 2.5 genes with 4 insertions, about half of the 5 genes found might be hotspots. Thus, there is a 50% chance that any of these genes are hotspots.

5. In this example, the X mutants are analogous to the *r*II mutants that Benzer worked with, and *E. coli* strain A is similar to Benzer's strain B, whereas *E. coli* strain B here is similar to Benzer's strain K.

To carry out crosses with different X mutants, we would coinfect *E. coli* strain A with two different X mutants and plate dilutions of the resulting lysate on strain B. Only the recombinants to wild type can grow, and the number of plaques minus the plaques resulting from the single parent controls is proportional to the recombination frequency between the two mutations in gene *X*.

6. (A) If there was an intron in the middle of the *r*II region, shown in the text as being in the middle of the *r*IIB gene, it would not have affected either (*a*) the order of genetic markers or (*b*) complementation with mutations in a second gene. There would be no complementation observed between the split segments of the *r*IIB gene. However, (*c*) two-factor-cross frequencies would be affected for any mutations separated by the intron, since it would increase the distance between them. (*d*) Point mutation hotspots would not be affected, but (*e*) deletion endpoints would be very frequent in the interval between the last mutation in the first part of *r*IIB and the first mutation in the second part of *r*IIB.

The apparently large deletion interval, together with the high recombination frequency for two mutations on either side of the intron boundary, could have suggested to Benzer that there was a segment of DNA in the middle of the gene that did not encode essential amino acids. None of Benzer's experiments could have determined whether it was a segment that encoded nonessential amino acids or simply spliced out DNA or RNA before the final message was translated.

(B) When the structure shown in the text is drawn out as a linear duplex, it is functionally equivalent to having an intron, as in Problem 6A. Therefore, the answers to Problem 6A apply here.

(C) A mutation is a base-pair change, and even though two amino acids in different parts of the protein might be affected by one mutation, the mutations map as a single point in the DNA. Therefore, (*a*) the order of mutations in the *r*II gene would not be affected. Likewise, (*b*) complementation occurs between gene products in the cytoplasm and would be unchanged by the gene structure shown in the text. In other words, complementation with a second *r*II gene would be the same as Benzer reported. In fact, (*c*) two-factor-cross frequencies, (*d*) definition of hotspots, and (*e*) occurrence of deletions would all remain the same.

(D) As in Problems 6A–C above, (*a*) the order of genetic markers would be the same. However, (*b*) complementation would now reveal two gene segments, which would appear as two cistrons, and (*c*) two-factor-cross frequencies would be greatly increased for any mutations in segment a in crosses with mutations in segment b. (*d*) The occurrence of hotspots would not be affected, but (*e*) many deletion endpoints would enter each gene from one end and still not affect the other gene segment. This would probably have indicated to Benzer that the gene segments were separated.

(E) As in Problems 6A–C above, (*a*) the order of genetic markers would not be affected, and (*b*) complementation tests would still indicate that the a and b segments are part of the same cistron. However, (*c*) two-factor-cross frequencies between any mutation in the a segment and any mutation in the b segment would be affected, since they would be increased because of the intervening DNA. (*d*) The hotspots would be unchanged, but (*e*) the occurrence of deletion endpoints would be elevated for the interval between the last mutation in the a segment and the first mutation in the b segment (after drawing out the structure as a linear duplex).

The increased two-factor-cross frequencies and the large deletion interval would indicate that there was extra DNA between segments of the gene, but Benzer would not have been able to distinguish between the structures in Problems 6A–C.

Unit 2 / The Nature of the Genetic Code

1. a. The amino acid sequence is:
   ASN GLN SER ARG TRP ARG SER LEU ILE LEU MET GLY.

   b. The new protein sequence is:
   LSY SER VAL GLU MET ALA LYS PHE ASN THR ASN GLY.

   c. The new protein sequence is:
   ILE SER ARG ASP GLY GLU VAL —.

   d. Only the sequence resulting from a deletion of the first base of the sequence (as in *c* above) would be altered, and only in an *ochre* suppressor strain, to add the additional amino acids after the termination point. If the suppressor inserted tyrosine, then we would expect to see -(TYR) TYR --- and then termination again. At a few percent, we would expect to see a further extension by suppression of the second *ochre* codon to add -(TYR) TRP -. The entire suppressed chain would then read:
   ILE SER ARG ASP GLY GLN VAL TYR TYR TYR TRP.

2. **(A)** Let's take the sequence
   A A U C C G A C U C.
   In an overlapping code, this would code for the following protein sequence:
   ASN ILE SER PRO ARG ASP THR LEU.
   If we delete a base, say the C at position 4 to give
   A A U C G A C U C,
   then this would code for:
   ASN ILE SER ARG ASP THR LEU.
   In other words, the same sequence except that we have deleted one amino acid.

   Let's take a second sequence
   A G U A U C C G A U.
   This codes for:
   SER VAL TYR ILE SER PRO ARG ASP.
   If we delete the third base, U, we get
   A G A U C C G A U,
   which codes for:
   ARG ASP ILE SER PRO ARG ASP.
   Here, we have also deleted one amino acid but we have also changed two others.

   In general, there is the loss of one amino acid and the possible change of up to two other amino acids.

   (B) If we add a G after the 4th base in the first sequence in Problem 2A, we get
   A A U C G C G A C U C,
   which codes for:
   ASN ILE SER ARG ALA ARG ASP THR LEU.
   Here, we have added one amino acid and changed one other amino acid.

   If we add a C after the 5th base of the second sequence in Problem 2A, we get
   A G U A U C C C G A U,
   which codes for:
   SER VAL TYR ILE SER PRO PRO ARG ASP.
   This results in the addition of one amino acid.

   In general, there is the addition of one amino acid and the possible change of up to two other amino acids.

   (C) In an overlapping code, there is no correct reading frame, since all frames are used. Therefore, there is no frameshift that results from the loss or addition of a base pair, the only result being the loss or addition of an amino acid and the possible change of up to two other amino acids. After that, the encoded protein sequence is normal.

3. The first step in constructing the double mutant is to doubly infect *E. coli* B with each single mutant. The lysate is then plated on *E. coli* K to select for the recombinants that contain both the +1 and −1 frameshift mutations, since these will give the wild-type phenotype. There are four possible progeny from this cross: both parental genotypes, the wild-type recombinant, and the desired double mutant. Only the wild type and the double mutant will give plaques on *E. coli* K. These should appear in about equal numbers. We should take a set of four or five of these and do backcrosses with the wild type to distinguish between the two possibilities. Here, the recombinant we are testing is crossed with wild type in *E. coli* B and the progeny plated on *E. coli* B, since we want to find *r*II recombinants that are separated from the double mutant. Each single mutant will have the *r*II phenotype and give large plaques on *E. coli* B. We will see *r*II plaques resulting from a backcross of each double mutant we test, but not from any wild types that we test. We can then test these plaques further by purifying them and crossing them against each parental *r*II mutant to verify that they are indeed one of the two *r*II parents. The cross is done in *E. coli* B and the progeny are plated on *E. coli* K to select for wild-type recombinants. The

recovered *r*II mutants from the first backcross should give wild-type recombinants with one parent but not with the other.

4. The first cross is a backcross with wild-type phage in *E. coli* B. There are four possible progeny: the two parents and recombinants that contain either one or the other of the two −1 frameshifts. Everything but the wild type is an *r*II mutant, so we have to plate on *E. coli B* and pick *r*II plaques. Most of these plaques will be the double frameshift mutant, but some will be a single −1 frameshift mutant, either $-_x$ or $-_y$. The next cross is a test cross against each reference single frameshift mutant, -1_x and -1_y. The cross is done in *E. coli* B and plated on *E. coli* K. The double frameshift mutant will fail to give wild-type recombinants with either of the single frameshift mutants. However, the separated out single frameshift mutants will each give wild-type recombinants with one of the two reference mutants but not with the other. Subsequent testing should reveal the presence of each of the two −1 frameshift mutations among the *r*II mutants resulting from the first backcross.

5. Although this problem appears similar to Problem 4, the difference is that we do not have any reference mutants for test crosses. The first step is a backcross against wild type in *E. coli* B, plating the progeny on *E. coli* B. We pick *r*II plaques and collect a series of them. We then cross them against each other in pairwise crosses in *E. coli* B, plating on *E. coli* K, and look for those pairs that give wild-type recombinants. If we can find pairs that give wild-type recombinants when crossed with each other, this shows that the original *r*II mutant contained two different *r*II mutations. We have to screen many different pairs of recovered mutants to find the correct one. To control for our having found a spontaneous *r*II mutant in the first test cross, we should finally verify that the pair of *r*II mutants that does give wild-type recombinants when crossed together will not give wild-type recombinants when crossed back to the original suspected double *r*II mutant.

6. (*a*) The first step is a backcross with wild type in *E. coli* B, plating on *E. coli* B. If the phage was really a double mutant, *r*II plaques should result some percentage of the time from this cross, due to recombination that generates an *r*II mutant with only one mutation in the *r*II gene. (*b*) After collecting a series of *r*II mutants from a cross like this, the recovered *r*II mutants should then be crossed against each other in *E. coli* B, plating on *E. coli* K, to see whether any pairs give wild-type recombinants, which would show that they were indeed different *r*II mutations.

Unit 3 / Decoding the *amber* and *ochre* Triplets

1. The experiments we would carry out would be analogous to those done by Brenner and coworkers to elucidate the *amber* and *ochre* codons.

   a. We look for reversion of UGA mutants to wild type by treating the phage with hydroxylamine and then passing through *E. coli* B and plating on *E. coli* K. We do not find any revertants and conclude that there is no G•C pair in the codon or, if there is, it is related by a transition to another nonsense codon.

   b. We then convert UGA mutants to *amber* or *ochre* by treating the phage with hydroxylamine and passing through *E. coli* B and plating on *E. coli* K containing either an *amber* or an *ochre* suppressor. We find that we can detect *ochre* mutants this way. Therefore, there is a G•C pair in the UGA codon. Moreover, we can repeat the experiment without passaging through *E. coli* B and find that we can again convert the UGA to an *ochre* mutant. Therefore, the C must be on the sense strand for the UGA codon, and the G must be part of the codon.

   c. We find that we can induce UGA from wild type with hydroxylamine. We then carry out the same experiment that Brenner and his coworkers did. We collect a large number of *r*II mutants and separate out those that are UGA by testing with suppressors. The experiment involves treating wild-type T4 with hydroxylamine and then dividing the treated phage into two parts. The first part is passaged through *E. coli* B and then plated on *E. coli* B. The second is passaged through *E. coli* K and then plated on *E. coli* B. We gather a sufficient number of mutants to average more than 10 mutations at each UGA site. We then use mapping with deletions and point mutations to separate them into specific sites. We then compare the results from each set and ask whether there are sites that are found in both the phage passaged through B and the phage passaged through K. We also ask whether there are sites that are found only after passage through B and not when the treated phage are passaged through K. We find examples of both kinds of sites, and this tells us that there are two G•C base pairs, oriented differently, in wild-type codons that when changed to A•U base pairs yield UGA mutants. Thus, the UGA codon has two different A•U base pairs oriented differently (UA).

   Since the third base pair has been identified as G, then the composition of the codon, without respect to order, is (UGA).

2. **(A)** Reversion to wild type involves an AUG codon being converted to an AUA codon. After treatment with hydroxylamine, the C is initially affected. In this case, the C is on the sense strand (recall that the antisense strand has the same sequence as the mRNA) and altering it with hydroxylamine would immediately result in a changed message that would make wild-type protein in *E. coli* K. Thus, we would be able to recover revertants.

   (B) Reversion to wild type involves changing a CCX to a UCX. Here, the C is on the antisense strand, and upon infection of *E. coli* K, the message is still that of the mutant *r*II. Thus, we cannot revert this by infecting *E. coli* K directly after hydroxylamine treatment.

3. We can determine the orientation of the G•C base pair in the *amber* codon by treating an *r*II *ochre* mutant with hydroxylamine and dividing into two parts. The first part is passaged through *E. coli* B and then plated on *E. coli* K with an *amber* suppressor. The second is either plated directly on *E. coli* K with an *amber* suppressor or first passaged through the suppressor-containing *E. coli* K before plating on it. The resulting wild-type-appearing plaques must then be tested for growth on *E. coli* K with and without an *amber* suppressor to distinguish between true revertants and the conversions to *amber*. If the *ochre* codon is (UAA) and the *amber* codon is (UAG), then the action of this mutagen will result in a change on the antisense strand, which does not change the message before DNA replication. The message made in the *E. coli* K *amber* suppressor strain will still contain the *ochre* codon (with respect to base-pairing properties) and thus not be recovered without passage first through *E. coli* B. Therefore, we would expect to see conversion to *amber* only in the set that was passaged through *E. coli* B. These results would tell us that the A is on the antisense strand in the *ochre* codon, and thus the *ochre* codon is (UAA) and the *amber* codon is (UAG).

4. a. A mutagen that works on T (and presumably can convert it to any base) will change ATG (AUG) to UUG (LEU), GUG (VAL), CUG (LEU), AAG (LYS), ACG (THR), and AGG (ARG).

b. After passage through *E. coli* B, all codons will be detected provided they result in a revertant phenotype.

After passage through *E. coli* K, only those codons that arise from a change on the sense strand will be detected. Since T is the affected base, we have to look for a T on the sense strand. Recall that the mRNA codon has the same sequence as the antisense strand. The complement is the sense strand. The AUG codon (or ATG on the DNA) has the first A on the antisense strand and a corresponding T on the sense strand. Therefore, changes at this first position to UUG (LEU), GUG (VAL), and CUG (LEU) will be detected after passage through *E. coli* K.

5. Here, we carry out the same set B and set K experiment that Brenner and co-workers did. We mutagenize T4 with hydroxylamine and separate the culture into two parts. With set B, we passage through *E. coli* B before plating on *E. coli* B and looking for *r*II mutants. With set K, we first passage through *E. coli* K before plating on *E. coli* B and looking for *r*II mutants. We then find the *r*II mutants that are suppressible by the GAU suppressor by showing that they can make plaques on *E. coli* K with the GAU suppressor but not on *E. coli* K without the suppressor. We gather a large collection of GAU mutants from each set and map them into sites. We compare the occurrences of GAU mutations at each site from set B with those from set K. We find that there are some sites that are well-represented in both collections and some sites that are well-represented in set B but not in set K. By finding these two different categories of sites we show that the wild-type codons that can go to GAU by hydroxylamine are of two types. In one case, the C is on the sense strand, and in the other case, it is on the antisense strand. Therefore, in the resulting GAU codon, one U is on the sense strand and one U is on the antisense strand. In other words, the codon has an A and a U. Thus, the composition of the codon is (GAU).

6. We first mutagenize T4 with the hydroxylamine-like mutagen and then divide the culture into two parts. With one part, set B, we passage through *E. coli* B before plating on *E. coli* B and looking for *r*II mutants. With the second part, set K, we first passage through *E. coli K* before plating on *E. coli* B and looking for *r*II mutants. We then find the *r*II mutants from each set that are suppressible by the UCG suppressor by showing that they cannot plate on *E. coli* K but do form plaques on *E. coli* K with the UCG suppressor. We then map the UCG mutations in each set and assign them to specific sites. By comparing the occurrences of mutations at each site in each set, we find that there are two classes of UCG sites. One class is well-represented in both set B and set K, whereas the other is well-represented in set B but not in set K. This shows that there are two types of wild-type codons that can be converted to the nonsense codon. One codon has a T on the sense strand and the other has a T on the antisense strand. Therefore, the resulting nonsense codon has a C on the sense strand and a C on the antisense strand. In other words, the nonsense codon has a C and a G (-CG).

7. a. The glycine to arginine change results from a GGG/A to AGA/G change. Here, the G that changes to A is on the antisense strand, so the C that is directly affected by hydroxylamine is on the sense strand. The mutant message will be produced immediately upon infection of the mutagenized phage and it will not allow growth in *E. coli* K since it gives the *r*II phenotype. Therefore, this mutant will not be recovered.

b. The histidine to tyrosine change results from a CAU/C to a UAU/C change. Here, the C is on the antisense strand. The mutagenized phage still makes a wild-type message when introduced into *E. coli* K, and thus this mutant will be recovered even after passage through *E. coli* K.

Unit 4 / The Search for Colinearity

1. The key to this problem is to understand that since efficiencies of transductions can vary from experiment to experiment, it is necessary to control for this variable if one is to obtain recombination frequencies that are a function of only the true recombination frequency. Thus, for example, we would make a Trp^- derivative of each His^- mutant and prepare P1 phage on a Trp^+ derivative of a second His^- mutant. The number of Trp^+ colonies are scored after P1 transduction, as are the number of His^+ recombinants. By dividing the number of His^+ recombinants by the number of Trp^+ recombinants, we are able to normalize the His^+ frequency and obtain relative values that should be the same from experiment to experiment.

2. a. As in Problem 1, we would prepare a Trp^- derivative of each Met^- mutant, prepare P1 lysates on Trp^+ derivatives of a second mutant, and score both the Met^+ colonies and the Trp^+ colonies. Dividing the number of Met^+ recombinants by the number of Trp^+ recombinants gives a normalized value that should be constant for each specific pair of mutants.

   b. We would use the nearby Arg^- marker to carry out three-factor crosses. We would prepare P1 lysates on an Arg^- Met^- mutant and transduce a second Met^- mutant that was Arg^+. We select for Met^+ recombinants on medium containing arginine, to allow Arg^- mutants to grow. For example, if the order of mutations is *arg*⁻ *metB1 metB2* and the donor phage was *arg*⁻ *metB1* and the recipient *arg*⁺ *metB2*, then one crossover must be between the two *metB* mutations and the second must be on the right side of the *metB2* mutation. Therefore, the recipient will not inherit the *arg*⁻ mutation from the donor most of the time. On the other hand, if the order is *arg*⁻ *metB2 metB1* and the donor is *arg*⁻ *metB1* and the recipient *arg*⁺ *metB2*, then although one crossover is still between the two *metB* mutations, the second is on the left side of the *metB2* mutation. Many of the resulting Met^+ recombinants will inherit the Arg^- marker, depending on how close it actually is. If it is 95% linked, then 95% of the Met^+ recombinants will now be Arg^-.

3. We can carry out three-factor crosses. One type of cross forces a crossover between the closest mutations, *met*⁻ and *his*⁻, and then scores for inheritance of the third mutation, *arg*⁻. A second method selects for the outside marker and follows the pattern of inheritance for the other two markers.

   a. The first type of three-factor cross is done exactly as described above for Problem 2b. Here, we have a P1 donor that is Arg^- Met^- His^+ and a recipient that is Arg^+ Met^+ His^-. By selecting for Met^+ His^+ in the presence of arginine (minimal glucose medium with only arginine), we are forcing a crossover between the *met* and *his* genes. If the order is *arg met his*, then the recipient would retain the Arg^+ character, since the second crossover is to the right of the *his* gene. If the order was *arg his met* for the cross outlined above, then the second crossover would be to the left of the *his* gene and would bring in the Arg^- marker with a frequency that depended on the linkage.

   b. We could use a recipient that was Arg^- Met^- His^- and a P1 donor that was Arg^+ Met^+ His^+ and select for Arg^+ on glucose minimal medium containing histidine and methionine.

   If the order was *arg met his*, then we would expect to see the following as the most frequent classes of recombinants:

   Arg^+ Met^- His^-
   Arg^+ Met^+ His^-
   Arg^+ Met^+ His^+

   But we would not expect to see the four-crossover class Arg^+ Met^- His^+.

   If the order was *arg his met*, then the four-crossover class would be Arg^+ His^- Met^+, which would be exceedingly rare.

4. a. Two-factor crosses are prepared by having a recipient that is $MetB^-$ and also carries one of the *xyl*⁻ mutations. The donor for each P1 cross is $MetB^+$ and carries a different *xyl*⁻ mutation. The transduction mixture is plated on minimal medium with xylose as the carbon source to select for Xyl^+ recombinants. The mixture is also plated on glucose minimal medium lacking methionine. The number of Met^+ transductants is used to normalize each cross. We thus divide the number of Xyl^+ recombinants by the number of Met^+ recombinants to allow us to compare numbers from different crosses.

   b. We can set up three-factor crosses using two different *xyl*⁻ mutations and a *proK*⁻ mutation.

Then we set up the crosses as in the preceding two problems. The P1 donor carries the *proK*$^-$ mutation and one of the two *xyl*$^-$ mutations, and the recipient is *proK*$^+$ and carries the other *xyl*$^-$ mutation. We select for Xyl$^+$ recombinants on xylose minimal medium containing proline. We then examine the Xyl$^+$ recombinants to see what proportion inherited the Pro$^-$ marker from the donor. Depending on the order of the mutations (see solutions to Problems 2b and 3), most of the recombinants will be either Pro$^-$ or Pro$^+$.

5. There are two ways to detect X$^-$ mutants after UV treatment. We can plate the mutagenized cells on indicator medium and look for blue colonies. However, that would be very laborious. A better way is to use the penicillin selection technique referred to in the readings. Mutagenized cells are shifted to medium without compound x and time is allowed for one or two generations of growth before penicillin is added, which kills growing cells. The penicillin is then washed out, and the surviving cells are plated on indicator medium. This technique, which can be repeated for several cycles, enriches for X$^-$ mutants under the above conditions.

To determine whether there are hotspots, we first use P1 crosses to map the mutations into deletion groups and then use pairwise crosses to separate the mutations in each deletion group into different sites. Each cross is plated on selective medium without the x-compound to allow selection for X$^+$ recombinants. The number of occurrences at each site are compared with the expectation deduced from the Poisson distribution. If there are many sites with only 1 or 2 occurrences, then these can be used to calculate the Poisson distribution that they are part of, exactly as Benzer did.

6. a. Increased generalized recombination will cause an increase in the number of Gal$^+$ cells after generalized transduction, since the recombinants are generated by generalized recombination. The majority of transductants obtained by specialized transduction occur by integration of the transducing phage using the λ integration system. Increased generalized recombination would not increase specialized transduction, since the percentage of transductants generated by recombination with the *gal* region is small.

b. Here, generalized transduction would be unaffected, but specialized transduction with λ derivatives would be affected since λ infects the cell via the *malB*-encoded receptors.

c. A deletion of the λ attachment site would greatly reduce specialized transduction involving λ derivatives but not affect P1-mediated generalized transduction.

d. Because P1 transduction involves injecting a bacterial DNA fragment into the cell, this fragment would be broken down in the mutant described here and generalized transduction would be reduced. However, since the λ transducing phage have an origin of replication and can circularize, they would not be broken down and specialized transduction would remain unaffected.

e. In a Rec$^-$ cell, generalized transduction would be eliminated, since it depends on general recombination. However, since specialized transduction utilizes the phage integration enzymes, it would be largely unaffected.

7. **(A)** Yanofsky and co-workers would have found that the order of mutations in the gene (assuming that only mutations in the exons are detected) and the protein were still the same (colinear), but there would be a disparity between the recombination frequencies and the distance between amino acids affected by different mutations. For example, two mutations separated by the intron would have a large recombination frequency between them and yet still involve amino acids that were very close in the protein. Yanofsky et al. would have noted this and realized that there probably was intervening DNA, but they would not have been able to discern that the message was spliced from the primary transcript.

Sarabhai, Brenner, and co-workers ordered nonsense mutations by crosses and then looked at the pattern of peptide fragments generated by different nonsense mutations. The peptide fragments would be unaffected. The only clue that they would have had would again be the large recombination frequencies between mutants that generated a set of peptide fragments that differed by only one fragment. Although they might have suspected intervening DNA, it seems that the mapping and amino acid sequencing done by Yanofsky et al. would have put them in a better position to recognize that there was intervening DNA.

(B) There are several consequences of this structure. First, the mutations on the DNA affect a base pair, and each base pair change, or mutation, will map in a single point in the gene. However, when one looks at the protein sequence, as Yanofsky and co-workers did, each mutation in the gene would end up affecting, for the most part, two amino acids that were separated in the protein. The earlier in the gene (the farther left on the diagram) the mutation, the more separated the amino acids that are altered. As the mutations map deeper and deeper in the gene, the amino acids affected get closer and closer. It is likely that Yanofsky et al. would have guessed this structure.

Sarabhai, Brenner, and co-workers also would have ordered mutations in the gene by genetic mapping. Recall that they identified nonsense mutations and then looked at the pattern of soluble peptide fragments produced. Either one of the changed codons in the mRNA could be the nonsense mutation, and it is unlikely that the other change would lead to the creation of another nonsense codon. Since the size of the fragment yielding the tryptic peptides depends on the position of the nonsense site, they would not get a consistent set of peptides that would show the map was colinear. In some cases, the first altered point in the mRNA would be the nonsense site, giving a short fragment, and in some cases, the second altered point would be the nonsense site, giving a longer fragment. They would see many deviations from colinearity. In some cases, the second mutation might be a missense that altered the mobility of a soluble tryptic peptide, and they might suspect that more than one amino acid was being altered, but it seems unlikely that they would have been able to deduce this structure without additional experiments.

(C) Yanofsky and co-workers would have noticed that mutations at the end of the first segment (a) recombined with mutations in the second segment (b) at a high rate, even though the affected amino acids were as close or closer than those altered by two mutations in the same segment. This would have suggested that the (a) and (b) portions of the gene were separated. In addition, the mutations in the (b) segment of the gene would not be colinear with respect to the altered amino acids. There would be an inverse relationship between the order of mutations in (b) with respect to (a), and with the resulting altered amino acids. This would suggest that the mRNA for the second segment was read in the opposite direction off of the other strand.

Sarabhai, Brenner, and co-workers might have noticed the difference in recombination frequencies between mutations in (a) and (b), but it is not certain. However, nonsense mutations in the (b) segment would give a hierarchy of peptide fragments that was the opposite of that predicted by mapping. They would therefore have recognized the inverted nature of the second part of the gene.

(D) If both groups of authors had carried out their studies on this *r*II protein, they would have noticed the difference in recombination frequencies for mutations in the (a) segment from those in the (b) segment and certainly realized that two parts of the gene were separated by a significant distance. Otherwise, the order of mutations in the gene would still be colinear with the order of the resulting amino acid changes in the case of Yanofsky and co-workers. However, Yanofsky et al. would not have been able to distinguish this structure from that of an intron. In other words, although they could recognize intervening DNA, they could not tell whether the intervening DNA is passed over or removed at the DNA, RNA, mRNA, or protein level.

Sarabhai, Brenner, and co-workers would find that their results were not consistent with a simple one gene–one protein colinear pattern. Nonsense mutations in the (a) portion of the *r*II protein would make the peptide fragments up to the point of the termination but then fail to make the fragments at the C-terminal end of the (a) protein fragment. This would prevent fusion of the (b) portion. One would detect the peptides in the (b) fragment, except the first one, which would be missing the portion encoded by the C-terminal end of the (a) segment. Thus, with fragments encoded by the second part of the gene present, the results would be confusing. However, with nonsense mutations in the second part of the gene, all of the fragments from the first part of the gene would be present, except for the one involving the C-terminal end if the (b) termination fragment cannot undergo fusion. The fragments from the second part of the protein would give a pattern consistent with colinearity. It is possible that Sarabhai, Brenner, and co-workers might have guessed that the protein was assembled by fusion of two parts.

(E) This structure is formally analogous to that of a single intron. The answer is therefore the same as that for part A of this problem.

Unit 5 / Conjugation and Genetic Recombination

1. The genetic map of the recipient, with *lac* at an arbitrary 0 min, is as follows (the entire map is 93 min):

   lac (0/93 min), *leu* (7 min), *pur* (9 min),
   thr (17 min), *ile* (25 min), *val* (40 min),
   ala (55 min), *met* (75 min), *cys* (85 min).

   Drawing the order of markers from 0 to 93 in a clockwise fashion, Hfr 1 has its origin at 88 min and transfers markers in a clockwise fashion. Hfr 2 has its origin at 22 min and transfers markers counterclockwise. Hfr 3 has its origin at 35 min and transfers markers in a clockwise fashion. Hfr 4 has its origin at 60 min and transfers markers counterclockwise.

2. The recipient described in *a*, which carries λ at its attachment site, will have repressor (immunity) in the cytoplasm, which will prevent zygotic induction. When the integrated λ prophage, which is near the *gal* region on the chromosome, enters the cell, the repressor will prevent it from entering the lytic phase and lysing the cell, preventing the growth of recombinants with markers that enter after λ. However, if the recipient described in *b*, which does not have λ integrated and thus no immunity, was used, then the cell would be killed by the lytic cycle of the entering λ prophage.

3. Sensitivity to rifampicin is dominant to resistance, and since *E. coli* growing in rich medium has two chromosome copies, there is a segregation lag of several generations before a pure recombinant can segregate out and a phenotypic lag before all of the RNA polymerases that are sensitive to rifampicin can be diluted out.

Unit 6 / Genetic Regulatory Mechanisms

1. This experiment is directly analogous to the PaJaMo experiment.

   a. If the *xyl* genes are under the negative control of a repressor, then we would expect to see synthesis of the xylose breakdown enzyme begin several minutes after mating, continue at the maximal rate for about an hour, and then shut off. This is because it takes about an hour for the buildup of a sufficient concentration of repressor in the cytoplasm to block further synthesis.

   b. Immediately after introduction of the *xyl* region into the recipient, there is no activator in the cytoplasm, so no enzyme can be made. It takes perhaps an hour for the concentration of the activator to build up to sufficient levels in the cytoplasm to allow the synthesis. We would see no synthesis for an hour and then synthesis at the maximal rate only in the presence of xylose. In the absence of xylose, which stimulates the activator, we would not see any synthesis at all.

2. All three of the mutations result in the Lac$^-$ phenotype. Only the I^S mutation is *trans*-dominant. Thus, if we introduce an F′ factor that is $I^+Z^+Y^+$, only the cells containing the I^S mutation would still be Lac$^-$. To distinguish the strain with the P^- mutation from the strain with the Z^- mutation, we introduce a Z^-Y^+ F′. The P^- mutation prevents expression of the linked Z and Y genes, and the resulting diploid would still be Lac$^-$. The Z^- mutation (missense) would still allow expression of the Y gene, so the diploid would be Lac$^+$.

3. From the first strain in the table, it is clear that M and N must equal I or O and Y, although not necessarily in that order. The expression of β-galactosidase in the absence of inducer identifies an I^- or an O^c, and the failure to make permease in the presence of inducer identifies a defective Y gene (Y^-).

   The second entry confirms the above and also confirms that K or L equals Z.

   The third entry tells us that L cannot be Z or Y and thus must be I or O. Therefore, $\boldsymbol{K = Z}$.

   The diploid in the fourth entry is inducible for β-galactosidase and constitutive for permease. The functional Z gene cannot be linked to an O^c and the functional Y gene is. Since $K = Z$, we note that $\boldsymbol{L = O}$.

   The last diploid has an inducible permease, so the Y gene must be linked to a good O. If $N = Y$, then the L^-N^+ region in the diploid would yield constitutive permease. Therefore, N cannot be Y, so $\boldsymbol{M = Y}$ and $\boldsymbol{N = I}$.

4. a. Promoter mutations are *cis*-dominant, affecting only those genes linked to it on the same DNA molecule.

   b. This altered repressor will still bind to the operator in the presence of IPTG, so it will appear to be an I^S, causing the Lac$^-$ phenotype. Because of a kinetic effect, it will have a partial constitutive effect in the absence of IPTG. This is due to the fact that the repressor normally finds new operators after each cell division by rapidly binding and falling off nonoperator sequences. The off-rates even for sequences that mimic the operator are very rapid. However, with the affinity for the operator increased, the off-rates for operator-like sequences are now increased, and in a percentage of the population, the repressor cannot find the true operator in time. This results in a partial constitutive phenotype in the absence of IPTG. In complementation tests with wild type, the partial constitutive effect will be nullified by the wild-type repressor, but the I^S character will remain.

   c. These are really I^S mutations and will be dominant in the presence of a wild-type *lac* region, unless the copy number of the wild type is increased significantly relative to the I^S, and then most I^S subunits will be entrapped in a tetramer of three other good (wild-type) subunits. In this case, the wild type will be partially dominant.

Unit 7 / In Vivo Genetic Engineering

1. The object is to select for strains in which the F′ *arg* is integrated near the attachment site for a temperate phage that can be induced to yield specialized transducing phage. We can select for strains that have integrated the F′ *arg* by having 0.1 M $MgSO_4$ in the medium, and we can select for strains with integrations into one of the receptor sites for one of the virulent phage. We note that the receptor site for the V2 virulent phage, V2*rec,* is near the attachment site for the temperate phage D, *attD*. Therefore, we use the strain with the deletion for the *arg* genes but carrying the F′ *arg* and select for Arg^+ colonies in the presence of 0.1 M $MgSO_4$ and the virulent phage V2. We can test surviving colonies to see whether they yield Arg^+ transductants after first lysogenizing with phage D and then heat-inducing the phage and using the lysate to transduce an Arg^- strain on minimal medium without arginine. Transductants are then heat-induced again and the lysates used to transduce the Arg^- strain to Arg^+. The second cycle should yield high transducing titers.

2. We would start with the strain deleted for the *attD* locus. We then infect the strain with the temperate phage D on plates seeded with a D phage derivative that is both C^- and Int^-, as well as with the virulent phage V3. The first two phage allow selection for lysogens, and the V3 phage allows selection for disruptions of the V3*rec* locus. Many of the surviving colonies should have the D phage integrated into the V3*rec* locus, which is near the *ala* gene. Heat-induction of the integrated prophage should yield a lysate that will transduce Ala^- colonies to Ala^+, and true transductants should yield high transducing lysates upon subsequent heat-induction.

3. We would need a strain that is deleted for the *cys* region, an F′ *cys* factor that carries no homology with the region of the chromosome near the *cys* deletion, and a temperature-sensitive derivative of the F′ *cys* that cannot replicate at high temperature.

   The important point is that the receptor site gene for the V1 phage needs to be as close as possible to the attachment site for TP1. Therefore, we should put them adjacent to one another anywhere in the 75- to 100-minute region. We then select for Cys^+ colonies at high temperature in the presence of V1 phage. We lysogenize with TP1 and then induce and test for transduction of the Cys^- strain.

4. a. Inheritance of markers both before and after the *met* region will be normal.

   b. Markers donated before the *met* region, such as *mal, ala,* and *arg,* will be inherited normally, but markers after the *met* region will enter after the λ prophage. In this cell, there is no immunity in the cytoplasm and the λ will enter the lytic cycle and kill the recipient. Thus, observed inheritance of markers donated after the *met* region will be sharply reduced.

Unit 8 / Structure-Function Relationships in Proteins

1. a. Even though both amino acids are nonpolar, some contacts in the interior are very specific, and it is possible that the valine cannot make all of the contacts of the bulkier isoleucine.

   b. It is possible that leucine makes a contact involved in subunit interactions and cannot be replaced with glutamine, even though it is on the exterior.

   c. Residues such as glutamic acid in the crevice might be part of the active site or involved in the architecture of the active site, and even closely related amino acids such as aspartic acid might not be allowed.

   d. Proline can cause distortions in the secondary structure by interrupting α helices. Even at positions on the exterior it may not be allowed.

   e. Putting a polar (and charged) residue in the interior of the protein will almost always destabilize the protein.

2. a. Many substitutions of one nonpolar residue for another are tolerated in the interior of a protein.

   b. Most substitutions on the exterior are tolerated, since in many cases the amino acid is making no contact other than with water.

   c. Aspartic acid is similar to glutamic acid and can often be substituted for it.

   d. Many substitutions on the exterior are tolerated. If the position is not in an α helix, then proline would not necessarily cause a significant distortion in the secondary structure.

   e. The precedent for this is hemoglobin Sogn, mentioned on page 302. Here, the replacement of a leucine by an arginine occurs at an internal position. The polar group can protrude from the surface in this case and might form a salt bridge with an aspartic acid residue. There are no clinical symptoms displayed by people carrying this hemoglobin, although there is an increased rate of denaturation at very high temperatures and the appearance of some free α chains, not very disastrous considering the nature of the replacement. The key is that the charged group can protrude from the surface. In cases where this is not possible, the results are disastrous.

3. We initially infer that residues normally in the interior are hydrophobic. Thus, A and C are nonpolar. Although a number of changes on the exterior of a protein are not tolerated, we might guess that a change to proline is the most frequent cause of a defective protein. Thus, we might guess that D = PRO. This means that F represents an amino acid that is one base change from PRO and also from one of the interior amino acids, such as TRP or ILE. We note that both TRP and PRO are one base change from the UCG codon for SER. Thus, let's assign A to UGG (TRP), F to SER (UCG), and D to Pro (CCG), and also C to ILE. We also note that B is probably another polar residue and must be one step away from UGG. We can assign B to either ARG, since CGG and AGG are arginine codons, or CYS, since UGU and UGC are cysteine codons. We still have G, E, and H unassigned and HIS, LYS, and either CYS or ARG unassigned. If we assign B to arginine, then we cannot assign G and E to any two of the three remaining amino acids, since neither CYS, HIS, nor LYS is connected by a single base change. Therefore, we assign B to CYS. This leaves the possibility of assigning the G in the crevice to HIS (CAU or CAC), a frequent active site residue, and E to ARG (CGX, or here CGU or CGC). (An alternative is to assign G to ARG and E to HIS, since arginine is also a frequent active site residue.) This leaves lysine to assign to H, which fits, since the C (ILE) is AUA, AUU, or AUC and LYS is AAA or AAG. Thus, the codon for C would be AUA and the codon for H would be AAA. We can then summarize as one of several possible solutions:

   A = TRP, B = CYS, C = ILE, D = PRO,
   E = ARG, F = SER, G = HIS, H = LYS.

Unit 9 / Mutagenesis

1. In analogy with the studies of Farabaugh et al. and Albertini et al., among others, we should (*a*) scan the sequence for tandem repeats at which frequent frameshift mutations would occur and (*b*) also for sequences that are repeated after intervening stretches of DNA. These sequences are targets for deletions. (*c*) Based on the work of Coulondre et al., we should also look for 5-methylcytosine sites, which in *E. coli* K12 would occur at CCAGG and CCTGG sequences.

(*a*) Inspection of the sequence shows that the largest repeat of a single base is five, which is not particularly "hot." However, we do note near the end of the fourth line the sequence CATCCATC CATCCA, which repeats the 4 bases CATC four and a half times. This would be a significant hotspot, since it can mispair 10 bases during slippage.

(*b*) In the middle of lines 8 and 9 we have the same 11-base sequence TCCAAGAATCT. This would be a hotspot for deletions, as would the 11-base sequence CTAGCTTTAGG, which begins with the fifth base of the sequence and is repeated at the end of the first line and the beginning of the second. In addition, the very first 9 bases, AAGGCTAGC, is repeated before the middle of line 4. This would also be a site of frequent deletions, as would the 9-base sequence GGTATA TAG, which begins near the start of line 2 and is repeated in the middle of line 7. The 9-base sequence ATCGGGATC appears near the beginning of line 3 and at the end of line 8 and the beginning of line 9. These would be frequent sites of deletion formation between the repeats.

(*c*) There is a CCTGG sequence in the middle of line 5. The second C on each strand of the double-stranded DNA will be methylated in *E. coli* K12 and will be a hotspot for the C→T transition.

2. a. The codons for HIS are CAU and CAC. The following base substitutions are involved:

Strain 1:
TYR (UAU/C) to HIS (CAU/C) = A•T→G•C

Strain 2:
ASN (AAU/C) to HIS (CAU/C) = A•T→C•G

Strain 3:
ASP (GAU/C) to HIS (CAU/C) = G•C→C•G

Strain 4:
LEU (CUX) to HIS (CAU/C) = A•T→T•A

Strain 5:
PRO (CCX) to HIS (CAU/C) = G•C→T•A

Strain 6:
ARG (CGX) to HIS (CAU/C) = G•C→A•T

b. We would start out with Strain 4 and mutagenize with a mutagen such as EMS. After outgrowth to allow expression of the mutant phenotypes, the mutagenized cells would be plated on minimal medium containing both glucose and maltose and tetrazolium dye. We would look for mutator colonies by screening for colonies that have more red papillae (Mal^+ microcolonies growing out of the surface of the main colony). These would be purified and tested for the frequency of Mal^+ mutants in overnight cultures.

3. a. The codons that can go from wild type to UGA are:

CGA, AGA, GGA, UCA,
UUA, UGU, UGC, UGG.

G•C→A•T = CGA, UGG
G•C→T•A = GGA, UGC
G•C→C•G = UCA
A•T→T•A = AGA, UGU
A•T→C•G = UUA

b. For the G•C→A•T change,
codons at positions 9, 11, and 23.
For the G•C→T•A change,
codons at positions 12 and 21.
For the G•C→C•G change,
codons at positions 13, 18, 24, and 29.
For the A•T→T•A change,
codons at positions 7, 16, 19, and 25.
For the A•T→C•G change,
codons at positions 4, 15, and 22.

c. Here, only the G•C→C•G category would show UGA mutations, divided among the sites at codons 13, 18, 24, and 29.

4. a. When A* is created directly in the DNA, the mispairing will generate A•T→T•A and A•T→C•G changes. When the precursor is oxidized, it can be incorporated into the DNA across from A, yielding A•T→T•A changes when it pairs with T at the next round of replication and A•T→G•C changes when it pairs with G at the next round of replication. It can also be incorporated across from G, yielding G•C→T•A changes when it pairs with T at the next round of replication and G•C→A•T changes when it pairs with A at the next round of replication. It can also be incorporated across from the normal T, where on subsequent rounds it will mispair as if it were generated directly in the DNA. Therefore, five changes, A•T→T•A, A•T→C•G, A•T→G•C, G•C→T•A, and G•C→A•T will result.

b. The MutA protein might be a glycosylase that removes A* directly from the DNA when across from T, leaving a gap that is repaired. The MutB protein might then be a glycosylase that removes the A or G from A*•A and A*•G mispairs. The MutC product might be a hydrolase that converts A*TP back to the monophosphate, preventing its incorporation into the DNA.

5. a. The codons that can go to BYD via a single base change are DYD, XYD, YYD, BBD, BDD, BXD, BYB, BYX, and BYY. The categories are:

B•D→X•Y = BDD
B•D→Y•X = BBD
B•D→D•B = DYD, BYB
X•Y→B•D = XYD, BYY
X•Y→D•B = YYD, BYX
X•Y→Y•X = BXY

b. DYD, at position 12; BYB at position 3.

c. YYD at position 17; BYX at position 14.

d. We should note that the YBX sequence on the other strand would appear as a YDX sequence on this strand, with the Y undergoing frequent mutation. Therefore, we should look for both YBX and YDX. Codons that are part of a YBX sequence and include the X are rare here, and only codon 4 falls into this category. There is one YDX sequence, and codon 7 includes the Y. Therefore, codons 4 and 7 are part of hotspots.

6. a. **Strain 1:** DXX to BXX = B•D→D•B
Strain 2: XXX to BXX = X•Y→B•D
Strain 3: YXX to BXX = X•Y→D•B
Strain 4: BBX or BXB to BXX = B•D→X•Y
Strain 5: BDX or BXD to BXX = B•D→Y•X
Strain 6: BYX or BXY to BXX = X•Y→Y•X

b. Mutagenized cells of any of the six strains would be plated on medium with glucose, arabinose, and MacConkey dyes. Mutator colonies would have more red papillae than normal. After purification, these mutators would be tested for their ability to stimulate the Ara^- to Ara^+ reversion in each of the six strains derived in part *a*.

7. a. If T* in the DNA mispairs with C, it will lead to T→G or A•T→C•G transversions. If T* in T*TP is misincorporated into the DNA by pairing with C, then it will cause G•C→T•A transversions.

b. Here, the MutJ protein might be a glycosylase that removes T* from the DNA when paired with A, allowing repair synthesis to restore the normal T. The MutK protein might be a glycosylase that removes C from a T*•C mispair, allowing repair synthesis to restore an A, putting the T*•A back into the pool to be repaired by the MutJ protein. The MutN protein might be a hydrolase that hydrolyzes T*TP back to the monophosphate, preventing its incorporation into DNA.

8. The problem is that there are two different deamination products of DAP. One of these is guanine, which will pair with cytosine. However, the cell cannot have a glycosylase that removes guanine from the DNA, so this deamination product cannot be repaired. Even removing guanine across from thymine would be dangerous, since G•T pairs are generated in several ways. The deamination of 5-methylcytosine generates a G•T pair. If the G is removed and an A put back by repair synthesis, that would fix the mutation. In addition, G•T mispairs are generated as replication errors and in some cases the G is the correct base, whereas in other cases the T is the correct base. Repairing without strand discrimination would fix half the replication errors as mutations. Thus, the cell has passed up the opportunity to utilize DAP.

Unit 10 / Transposable Elements

1. The original *lacZ* mutation probably arose by a transposon such as Tn*10* inserting into the *lac* genes. This rendered the cells both Lac^- and Tet^r. Reversion to Lac^+ involves the precise excision of the transposon, resulting in the Tet^s phenotype.

2. This event probably occurs by the generation of a cointegrate, involving the fusion of the two plasmid circles into one plasmid circle, probably with one copy of the Tn*400* at each end of the cointegrate. The most likely explanation is that Tn*400* transposes by replicating a copy into the target sequence. In the mutant, the second step, the resolution of the cointegrate by site-specific recombination, is blocked.

3. Two λ phage carrying Tn*10*, one containing the wild-type *gal* region and the other with a *gal*⁻ mutation, are used to generate a heteroduplex phage with one strand containing the wild-type *gal* region and the other with the *gal*⁻ mutation. The phage cannot replicate its DNA due to mutations in the replication genes. This is used to infect a *gal* deletion strain, and the cells are plated directly on galactose MacConkey medium with tetracycline. The object is to determine whether there is a high proportion of sectored red and white colonies among the Tet^r colonies, indicating that heteroduplex DNA was transposed into the chromosome and suggesting that transposition is conservative, preserving both original strands of the DNA from the donor to the recipient.

4. We would prepare a Ty derivative containing an intron in the transposon. We would add glycerol to activate transcription and stimulate the transposition of the marked Ty element and then select for Kom^r cells. We then analyze the DNA of the *kup* gene in a set of Kom^r mutants. We not only expect to find the Ty DNA in the *kup* gene, but also expect it to now have lost the intron, which would be spliced out if transposition occurred via an RNA intermediate.

Solutions to Scenarios

Scenario 1 / Chances Are It's Poisson

A. The first step in solving problems of this type is to determine what events are being distributed into what intervals. Or, in analogy with Benzer's work, which are the "mutations" and which are the "sites." Here, we have a group of sky divers jumping into target squares. Clearly, the divers are being distributed among the target squares. Since we do not know the total number of divers or the number of squares, we cannot immediately calculate m for the distribution. However, we can use the ratio of two points on the distribution to solve for m and then to find any point and also the total.

Recall that $P_n = m^n e^{-m}/n!$ and, from the equations on page 534, the ratio of any two adjacent points $P_n/P_{n-1} = m/n$. In other words, $P_4/P_3 = m/4$.

The number of squares with four divers (which equals the total × P_4) is 2, and the number of squares with three divers (which equals the total × P_3) is 4. Thus, the total × P_4/the total × P_3 = P_4/P_3 = 2/4 = $m/4$ (from above). Therefore, $m = 2$.

1. There are several ways that we can calculate the number of squares with zero divers. Knowing m, we can calculate P_0 from the equation $P_0 = e^{-m}$ and then multiply that by the total, which we must first calculate (see below). Alternatively, we can take ratios. We can either calculate the number of squares with two, one, and zero divers in steps or take the ratio P_3 × the total/P_0 × the total = squares with three divers/squares with zero divers = $P_3/P_0 = m^3/3!$ = 8/6 = 1.33 = 4/squares with zero divers and, rearranging, **the number of squares with zero divers = 4/1.33 = 3.**

2. To get the total number of squares, we can calculate any of the points, such as P_3 or preferably P_0, which we now have. Thus, **the number of squares with zero divers = total squares × P_0 = 3** and, rearranging, the total = $3/P_0 = 3/e^{-m} = 3/e^{-2}$ = 3/.1353 = 22. Therefore, the total number of squares is **22,** and since $m = 2$, the total number of divers is **44**.

B. Once again, it is important to recognize which events are being distributed among which intervals. In this problem, 200 darts are thrown at a dart board with 400 intervals, each interval corresponding to one stock. Thus, the darts are being distributed among the stocks. Therefore, m = **200/400 = 0.5.** Here, we should calculate the expected Poisson distribution for $m = 0.5$, assuming that the distribution is completely random. We can then see whether the number of darts landing on one stock is very much more than expected. Since we have the total, 400, it is easy to calculate each value directly by simply plugging values into the Poisson equations. Thus, $P_0 = e^{-m}$. From the values given on page 600 of the text for this problem, $e^{-m} = 0.6$ when $m = 0.5$. To find the number of stocks on the dart board that will have no darts hitting them, we multiply the total (400) by 0.6 and get **240**. We can then calculate the number of stocks that are expected to have 1, 2, 3, and 4 darts each by either using the ratio method or just calculating directly from each equation. The obtained values are:

Number of darts	*Expected number of stocks*
0	240
1	120
2	30
3	5
4	0.62

We can see that it is improbable that more than one stock has 4 darts. Since there are five stocks with 4 darts, we expect that four of these five are "hotspots" and not part of the random distribution, in direct analogy with Benzer's determination of hotspots. (The total does not add up to exactly 400 because here we used the rounded off value on page 600 for e^{-m} instead of the exact value.)

C. 1. To calculate m for this distribution, we have to recognize that the intervals are the 36 numbers and the roulette ball is being distributed among them at each spin. So, there are 360 spins and the roulette ball falls into one of the 36 intervals each time. Therefore, $m = 360/36 = 10$.

2. By solving the Poisson equation for $m = 10$ and $n = 10$, we get $P_{10} = 0.125$. But, we now have to multiply this by the total number of intervals, or numbers in this case. Why? Because P_{10} is the probability that any **one** specific number will appear 10 times. There are 36 numbers, so if we want to know the expected number of occurrences that **any** number will appear 10 times, we have to multiply by the total number, or by 36: **0.125 x 36 = 4.5.**

3. By solving the Poisson equation for $m = 10$ and $n = 21$, we get $P_{21} = 0.000886$, and multiplying this by 36 we get **0.032**. Now let's consider what this number represents. Even though it is the expected number of times that some number will appear 21 times out of 360 spins, when this expectation is less than 1.0 it can be seen as a probability in its own right. Namely, if we went to the casino and monitored 360 spins of the wheel, there would be a 0.032 chance (about 1 chance in 31) that one of the numbers would come up 21 times. Alternatively, if we visited the casino every night for 31 days, we would expect to see this happen on one of those nights. Because the problem stated an arbitrary measure of improbability of 0.02 (1 out of 50), we cannot say that this roulette wheel is defective.

D. Since there are 12 horses in each race, there are 144 possible daily double combinations.

1. In two years, 710 daily doubles were run. Each time a daily double set of races is run, the winning combination will be distributed over one of the 144 possible combinations. Therefore, here, the **daily double races are the events**, and the **144 combinations are the intervals** into which the events are being distributed. We can then calculate m for this distribution by taking 710 and dividing it by 144. Thus, **710/144 = 4.93 = m.**

To calculate the expected number of times that some combination would come up 14 times, we first use the Poisson formula to calculate P_{14} for $m = 4.93$, which gives **0.000415** (or 4.15×10^{-4}). Now, P_{14} **is the probability or fraction** of combinations that will come up 14 times out of 710 trials. We need to multiply this by the total number of combinations (144) to find the **expected number** of times that a combination will appear 14 times: **0.000415 x 144 = 0.06**. Therefore, the expectation of some combination coming up 14 times is 0.06 (close to one out of sixteen). Since this is greater than the arbitrary value we set of 0.001, then we cannot say that the races are dishonest.

2. This is solved in the same manner as Problem D1 above. Here, $m = 720/144 = 5$. $P_{18} = 4.0146 \times 10^{-6}$. Multiplying this by 144 gives **0.00058**, which is less than 0.001. In this case, we might assume that the races are fixed.

3. Here, $m = 700/144 = 4.86$. $P_{17} = 1.0246 \times 10^{-5}$. Multiplying this by 144 gives **0.00148**, which is higher than 0.001, so we cannot say that the races are dishonest.

E. 1. There are 52 different cards, and you have 300 draws. Each draw could be any of the 52 cards. Therefore, the **drawn (or selected) card is the event** that is being distributed among one of **52 intervals.** To find m, we take 300 and divide it by 52. Thus, $m = 300/52 = 5.769$. To find the expected number of cards that will appear 18 times out of 300 draws, we first calculate the probability or fraction of cards that will appear 18 times and then multiply this fraction by the total number of cards (52). P_{18} for $m = 5.769 = 2.443 \times 10^{-5}$. Multiplying this by 52 gives **0.00127**, the expected number of occurrences of any card appearing 18 times. Since this is just higher than our arbitrary value of 0.001, we cannot say that the machine is cheating us.

2. In this case, m **= 200/52 = 3.846.** P_{15} **= 9.735 x 10^{-6}**. Multiplying this by 52 gives **0.0005**, the expected number of occurrences of any card appearing 15 times in this sample size. Since this is lower than 0.001, we can say that the machine is cheating us.

F. Each number picked can be one of 49 possible numbers. The **picked numbers are the events**, and the **49 numbers are the intervals** into which they are distributed. To calculate m, we therefore take 730 and divide it by 49. Thus, m **= 730/49 = 14.898.**

1. To see the expected number of times that any number comes up either 1 or 31 times, we first calculate the probability or fraction of numbers that come up either 1 or 31 times and then multiply it by the total number of numbers, or by 49. Therefore, for m = 14.898, P_1 **= 5.0467 x 10^{-6}**. Multiplying this by 49 gives **0.000247.** Since this is less than the arbitrary value of 0.001, we can say that the lottery is fixed based on this result. In addition, for m = 14.898, P_{31} **= 9.59 x 10^{-5}**. Multiplying this by 49 gives **0.0047**, which is greater than 0.001. Therefore, this result alone would not allow us to say that the lottery is fixed.

2. In this case, for m = 14.898 as above, P_3 **= 1.866 x 10^{-4}**. Multiplying this by 49 gives **0.0091**, which is higher than 0.001, so this result alone does not allow us to say that the lottery is fixed. In addition, P_{35} **= 3.759 x 10^{-6}**. Multiplying this by 49 gives **0.000184** (or 1.84 x 10^{-4}), which is less than 0.001, meaning that we can say that the lottery is fixed.

3. In this case, for m = 14.898 as above, P_2 **= 3.759 x 10^{-5}**. Multiplying this by 49 = **0.00184**, which is greater than 0.001. In addition, P_{34} **= 8.832 x 10^{-6}**, and multiplying this by 49 gives **0.0004327** (or 4.327 x 10^{-4}), which is less than 0.001, allowing us to say that the lottery is fixed.

G. There are 365 days or dates in the year. Each student has a birthday that can fall on one of the 365 days. Therefore, we are distributing the students birthdates among the 365 days. There are thus **365 intervals** into which the students' birthdates are being distributed. We can thus calculate m by dividing the number of students by 365.

1. In this case, m **= 26/365 = 0.0712.** There are two ways to find the number of students with birthdays on the same dates. One is to first calculate the number of students with unique birthdates and then subtract this from the total number of students, the remainder being the desired number. To do this, we must really calculate the number of days with only one student's birthdate. We do this by finding P_1, the fraction of days with one student, and multiplying it by the total number of days, or 365.

   For m = 0.0712, P_1 **= 0.0663.** Multiplying this by 365 gives 24.2, which is rounded off to **24.** Subtracting 24 from 26 gives the expected number of **2** students with birthdays on the same date as one other student. We can also calculate the number of days with more than one student. In this case, P_2 **= 0.00236.** Multiplying this by 365 = **0.86.** Note that this is the expected number of days with two students birthdays falling on it. The number of students is by definition twice this, which is **1.72**, rounded off to **2.** This is the same result as we calculated before.

2. In this case, m **= 34/365 = 0.093.** P_1 **= 0.0847.** Multiplying this by 365 gives 30.9, or **31.** Subtracting 31 from 34 gives the expected number of **3** students with birthdays on the same date as one other student. Alternatively, from P_2 **= 3.94 x 10^{-3} x 365 = 1.44**, we get the number of days with two students. Multiplying this by 2 we get the number of students with birthdays on the same date as one other student, which is 2.88, rounded off to **3.**

3. In this case, m **= 73/365 = .2.** Here, P_1 **= 0.1637.** Multiplying this by 365 gives 59.8, rounded off to **60.** This is the number of days with one student, and thus the number of students with birthdays on unique days. To find the expected number of students who share birthdates, we subtract 60 from 73, to get **13.** However, this is not the complete story. Let's look at the expected number of days with two students and even with three students to see how those 13 students are themselves distributed. P_2 **= 0.01637.** Multiplying this by 365, we get 5.97, or **6.** This is the number of days with two students, so we have to multiply this by 2 to get the number of students who have birthdays on the same day as one other student, which equals **12** students. P_3 **= 0.00109.** Multiplying by 365 we get 0.398 or **0.4.** This is the expected number of days with three students. To find the expected number of students who have their birthdays on the same day as two other students, we multiply by 3 to get **1.2.** Note how

the total number of students who share birthdates still comes out to 13 (12 + 1.2 = 13.2, rounded off to **13**). In reality, this last category can only be occupied by 0, 3, 6, etc., students. Because the expected number of days with three students is a little more than 1/3, with this sample size we would expect three students to have a birthdate on exactly the same day about 1/3 of the time and we would not expect to find any about 2/3 of the time.

H. In this problem it is important to recognize what is being distributed into which intervals. Each time the wheel spins, one of the intervals on the wheel is selected. Therefore, the spins are being distributed among the 80 intervals of the wheel.

1. If we want to look at the $100,000 prize, we note that there are **4 slots** on the wheel for this value. Suppose we grouped them all together. These 4 slots would represent **4/80** of the wheel, or 1/20 of the wheel. We can then conceive of the wheel as having 20 slots or intervals of equal length or probability as the combined $100,000 slots. Therefore, we are really distributing the results of 1120 spins over 20 slots or intervals, and m **= 1120/20 = 56**.

   We can now proceed to calculate the expected number of times that the $100,000 prize comes up as few as 24 times out of 1120 spins. For m = 56, $P_{24} = 7 \times 10^{-7}$, a very low number. This is the probability that any one slot (of the length of four of the original slots) would come up only 24 times out of 1120 spins. Now in most of the problems we have been considering, we would multiply this by the total number of slots, which is 20 in this case. However, in this specific problem, we start out with the suspicion that the higher number slots are fixed. So, in reality, we are interested in how the $100,000 slot does. In this case, we would want to calculate the probability that this specific slot or set of slots is underrepresented, and that means that if we multiply by 20, we should really divide by 20 again. This is one of those situations where the probability of any one number is what we are after. Of course, in this example, the probability is so low that it really doesn't matter, since any way you look at it, this Big Spin is fixed.

   To calculate how often we would expect the $1,000,000 prize to come up only 12 times, we note that there are 2 slots with a value of $1,000,000. So, if we place these together, they represent a combined slot or interval that is 1/40 of the wheel. Therefore, we really consider the results of 1120 spins being distributed into 1 of 40 slots, one of which is the $1,000,000 slot. For this distribution, m = **1120/40 = 28.** Here, $P_{12} = 0.000335$. This is the probability that the $1,000,000 value will come up only 12 times out of 1120 spins. If we wanted to know the expected number of times that any of the equivalent slots would come up only 12 times, then we would multiply this number by 40. However, we suspect that the high values are being preferentially underrepresented, so we want to know the expected number of times that the $1,000,000 slot itself appears only 12 times. (If we had multiplied by 40, we should divide by 40, to give the expectation for a single number.)

2. In this case, as above, for the $100,000 slots, m **= 1120/20 = 56.** $P_{20} = 1.8 \times 10^{-8}$. This is the probability that any one number will come up only 20 times, and thus becomes the expected number of times that the specific $100,000 slot will only come up 20 times out of 1120 spins. For the $1,000,000 prize, as above, m = **1120/40 = 28.** $P_{14} = 0.00144$. As above, we do not multiply this by 40, since we are interested in the probability of one specific interval, namely, the $1,000,000 slot.

3. As above, we calculate m for the distribution concerning the $100,000 slot by taking 1200 and dividing by 20. Thus, m **= 60**. For this distribution, $P_{40} = 0.0014$, which is the probability that any one slot will come up as few as 40 times. It is also the expected number of times that the specific $100,000 slot will come up as rarely as 40 times. For the $1,000,000 slot, m = **1200/40 = 30.** $P_{16} = 0.0019$.

4. As above, we calculate m for the distribution concerning the $100,000 slot by taking 1360 and dividing by 20. Thus, m **= 68**. For this distribution, $P_{36} = 7.37 \times 10^{-6}$, which is the probability that any one slot will come up 36 times. It is also the expected number of times that the specific $100,000 slot will come up as rarely as 36 times. For the $1,000,000 slot, m **= 1360/40 = 34.** $P_{14} = 5.4 \times 10^{-5}$.

Scenario 2 / Munchkin Genetics

A. 1. The key to the method used by Beckwith and Signer is to select for strains in which an F′ factor carrying the gene to be cloned has integrated into the chromosome near the attachment site for a temperate phage. In this problem, the element similar to the F′ factor is the RRS element. Moreover, one has an RRS′ *crg* element that is temperature-sensitive (for replication). We note that the WWE-*rec* locus is near the *att-munchk* site on the *Escherichia yellow brick* chromosome. Therefore, since we have selective medium for each marker, if we use the strain deleted for the *crg* gene and containing the temperature-sensitive RRS′ *crg* element, we would plate cells at high temperature, on medium selecting for Crg^+, in the presence of WWE virulent phage. Only cells that have integrated the RRS′ *crg* element into the chromosome can grow on Crg^+ selective medium at high temperature. Only those cells with a mutation in or disruption of the WWE-*rec* locus can survive exposure to the WWE phage, since they will have a defective receptor. Cells in which the RRS′ *crg* element had integrated into the WWE-*rec* locus will survive the phage exposure. Therefore, survivors of this plating should have the RRS′ *crg* element integrated into the WWE-*rec* locus.

Because the WWE-*rec* locus is near the *att-munchk* site, we would expect that a small fraction of the phage that excise from this locus might have neighboring genes attached. We would first lysogenize the cells with the *att-munchk* phage by infecting with the *att-munchk* phage and testing survivors for immunity and inducibility of the phage. Lysogens are then induced with UV (ultraviolet light), and the lysates are used to transduce a strain deleted for the *crg* gene. Cells are plated on medium selecting for Crg^+. The rare Crg^+ colonies are then purified and induced with UV, and these new lysates are again used to transduce a strain deleted for the *crg* gene. Now, a high frequency of transduction should be observed, since this second-stage lysate is highly enriched for phage carrying the *crg* gene.

2. The transpositions are generated by the integration of the RRS′ *crg* factor, which creates an Hfr strain. Here, we would use the transposed strain as a donor in an Hfr cross to a multiply marked recipient strain and test for which markers were donated efficiently in a short mating, or else do a classic interrupted mating experiment. We would see that either the *tin man* locus was donated early and the *scarecrow* locus late or vice versa, depending on the orientation of the integrated RRS′ *crg*. In principle, the interrupted mating experiment would show either the order *tin man - emerald city - Oz - poppy fields - lollipop guild - kansas - wizard - scarecrow* or the reverse order.

B. The key point here is that in the second cross we are witnessing the equivalent of zygotic induction. Here, the direction of transfer must be *wizard - scarecrow - att-munchk - tin man*, etc. When the integrated *att-munchk* phage is transferred into a cell that does not carry the *att-munchk* phage and thus has no immunity in the cytoplasm, the prophage is induced and activates virulent functions that allow it to enter the lytic cycle, killing the cell. Markers that enter before the prophage are seen to be inherited in surviving colonies, because in many cases these markers enter on a chromosomal fragment that has broken off before the *att-munchk* phage is transferred. Markers that enter the cell after the integrated phage are rarely inherited, because cells receiving a chromosomal fragment with these genes also receive the integrated phage, which is then induced.

Scenario 3 / Cornering the Market with *Escherichia Monopoli*

A. 1. The best way to isolate a mutant with a deletion of the *Pennsylvania Avenue* gene is to use the concept described in the Beckwith et al. (1966) paper for finding deletions that extend into the *trp* region. The point is to select for resistance to infection by a virulent phage that depends on receptors encoded by a gene that is near the locus of interest. A proportion of spontaneous phage-resistant mutants will have deletions extending into neighboring genes. The resistant mutants can be screened for simultaneous loss of the function of interest. In the case of *E. monopoli,* we note that the *Pennsylvania Avenue* gene is directly adjacent to the *Shortline Railroad* locus, which encodes the receptors for the Shortline Railroad phage. So, we would select Shortline phage-resistant mutants by infecting the *E. monopoli* cells with the phage and then looking for surviving colonies. After purification, these colonies would be tested for the ability to grow on presidential medium, which requires the Pennsylvania Avenue$^+$ phenotype. Strains that are phage-resistant and cannot grow on presidential medium would have deletions that extend from the Shortline-*rec* gene into the *Pennsylvania Avenue* gene.

2. Here, we would start with the strain we constructed in part 1 above. We would introduce into this deletion strain the temperature-sensitive F′ factor that carries the *Pennsylvania Avenue* gene. We could cross this into the deletion strain, selecting for strains that are Shortline phage-resistant and can grow on presidential medium. Note that the deletion strain is phage-resistant and the phage-sensitive donor can transfer the F′ factor, which confers the ability to grow on presidential medium.

We would then infect the strain with an excess of Reading Railroad phage and plate on presidential medium at high temperature (presumably 42°C). We use the Reading Railroad phage because its receptor gene is adjacent to the integration site for the temperate phage Income Tax. Mutants that can grow on presidential medium at high temperature must have the F′ *Pennsylvania Avenue* factor integrated in the chromosome, and mutants that can survive the phage infection must have an inactivation of the Reading Railroad-*rec* locus. Cells in which the F′ factor integrated into the Reading Railroad-*rec* locus will have both properties. These can be tested to see whether they are Hfrs with points of origin at the Reading Railroad-*rec* locus.

The next step is to infect the strains with the Income Tax phage. Many of the survivors will be lysogens that now carry the Income Tax phage at its integration site. These are then induced with UV, and the lysates are used to transduce the starting strain that is deleted for the *Pennsylvania Avenue* locus, the transduction mix being plated on presidential medium. Transductant colonies are then used to make a new lysate, which should now have a high proportion of transducing particles and should transduce the deletion strain to Pennsylvania Avenue$^+$ at a high frequency.

B. To use the method employed by Shimada et al., we first have to delete the normal attachment site for the Water Works phage. Therefore, we need to start with the strain that has a deletion extending from *Marvin Gardens* to *Atlantic Avenue,* since this deletes the attachment site. We then use a temperature-inducible mutant of the Water Works phage and infect the starting strain. We select for rare lysogens by plating on medium seeded with **Water Works phage mutants** that have **no immunity** and that **cannot integrate** into the chromosome. The survivors will be either mutants that are resistant to infection by the Water Works phage due to a mutation in the phage receptor site or lysogens in which the Water Works phage has integrated at a **secondary attachment** site. Resistant mutants will also be resistant to infection and killing by a virulent mutant of the Water Works phage. The lysogens will form colonies at low temperature (30–34°C) but not at high temperature (42°C), since the prophage will be induced at the higher temperature. We also will have added **Pennsylvania Railroad phage** to the plates to select those rare lysogens in which the Water Works phage had integrated into the **Pennsylvania Railroad-*rec*** gene, since this gene is next to the *St. James Place* gene that we want to clone on a specialized transducing phage.

We then **heat-induce the prophage** and use the lysate to transduce the strain carrying the deletion extending from *New York Avenue* to *St. James Place,* **selecting** rare **St. James Place$^+$ transductants** that grow on elitist medium. Colonies that grow on this medium are likely to be transductants carrying the *St. James Place* gene. These are then in turn heat-induced and the lysates again used to transduce the same deletion strain to St. James Place$^+$ on elitist medium. The frequency of St. James Place$^+$ transductants should be high for those lysates made from colonies that contained a specialized transducing phage carrying the *St. James Place* gene.

C. The important concept here is that since the recipient strain does not contain the Water Works phage, there is no immunity in the cytoplasm, and transferring in the Water Works prophage during an Hfr cross will result in **zygotic induction**. Markers that are transferred before the Water Works phage will be inherited at their normal rate, since most of the markers are part of transferred fragments that have broken off before the later markers are transferred. Since the integrated Water Works phage is transferred after *Pennsylvania Avenue* and *Pacific Avenue*, these markers will be inherited normally. However, *Atlantic Avenue* and *New York Avenue* are transferred after the integrated Water Works phage. Every donor chromosomal fragment carrying these two loci will also carry the Water Works phage, which is induced upon entry into the nonimmune recipient, resulting in cell lysis.

The surviving colonies that inherit the *Atlantic Avenue* or *New York Avenue* gene are therefore sharply reduced.

Scenario 4 / Cloning à la Bogie

A. For the *Maltese Falcon:*

1. Using the same concept as in the Scenario 3, we can generate a deletion by **selecting for resistance to a virulent phage** that uses a receptor site encoded by a gene very near the locus we want to delete. By consulting the map of *Escherichia bogie,* we note that *Maltese Falcon* is right next to Señor Ferrari-*rec,* the receptor site gene for the Señor Ferrari virulent phage. Therefore, we select colonies resistant to this phage and then test them for the inability to produce endorphin (colonies that fail to turn red in the presence of a 340-nm frequency laser beam). Those that are simultaneously resistant to the phage and Maltese Falcon$^-$ must have deletions of the *Maltese Falcon* gene.

2. Isolating an F′ factor was not directly covered in the papers reproduced in the text. However, from the cited references, or from thinking about it, we can devise a method for selecting rare derivatives of F that carry the gene of interest. In the case of the *Maltese Falcon* gene, we would first **select Hfr strains that have integrated F next to the gene** and then find F′ derivatives that have excised aberrantly from the chromosome and now carry the *Maltese Falcon* gene. We can use the principles of Beckwith and Signer to select strains in which F has integrated into the Señor Ferrari-*rec* locus by, in this case, selecting colonies at 42°C that can grow on medium seeded with Señor Ferrari phage. The surviving colonies are tested to show that they are indeed an Hfr with the origin of replication at the Señor Ferrari-*rec* locus by using the strains with point mutations in different genes (see p. 608) as recipients. Here, we would plate the mating mixture on medium selecting first for **Sahara$^+$** and then **Virginia City$^+$**, since these are the markers donated first or last by the desired Hfr depending on the orientation. Only a fraction of the cells might be the desired Hfr, since a single mutation in the chromosome would give the selected phage resistance, but one might expect as many as 1 out of 10 of the survivors to be the correct Hfr. (It is also possible to select for cells that maintain the F factor by using male-specific phage such as R17 or f2 and requiring an integration event of F in the chromosome. This would constitute a double selection.)

How do we take the above Hfr strain and find an F′ factor carrying the *Maltese Falcon* gene? Recall that among the tools we are provided with is a Rec$^-$ strain that cannot undergo normal recombination. We can use this strain to isolate a deletion of the *Maltese Falcon* gene by the procedure described in part 1 above and then employ it as a recipient in a cross with the Hfr we have generated. Here, if we had a selection, then we could **select for Maltese Falcon$^+$ strains** that can grow on unsupplemented medium. Since the **Rec$^-$ recipient cannot incorporate donated fragments by recombination**, the Maltese Falcon$^+$ recipients that are generated from this cross result from those strains that receive an F′ factor carrying the *Maltese Falcon* gene. However, in this problem, we are not given a direct selection, but rather a screening method. So, we would have to screen for laser-induced red colonies after a cross in which we ensure that the donor strains do not survive (either by having the recipient Str^r and using streptomycin, or by making the recipient resistant to one of the virulent phage and infecting with that virulent phage as a way of killing the donor after the cross). These colonies would be very rare, but, when purified, they yield a high frequency of transfer of the Maltese Falcon$^+$ character to a similar recipient in a subsequent cross. Alternatively, we could select for F′ derivatives that carry the *Sahara* gene, since we are provided with Sahara$^-$ strains and even Rec$^-$ Sahara$^-$ strains. We would construct a *Maltese Falcon* deletion derivative of the Rec$^-$ Sahara$^-$ strain and then directly select for F′ factors that result from aberrant excision and have incorporated at least the *Sahara* gene. These would be screened for laser-induced red colonies to indicate those that have also incorporated the *Maltese Falcon* gene into the F′ factor.

Cloning the *Maltese Falcon* Gene: Using the tools constructed above, we would start out with a strain that is **deleted for the *Maltese Falcon* locus** and that **carries the F′ *Maltese Falcon* derivative** that cannot replicate at high temperature. We need to select rare strains in which the F′ factor has integrated into a locus near enough to a temperate phage attachment site so that we can later select transducing phage carrying the *Maltese Falcon* gene. In this problem, the gene into which the F′ integrates must be adjacent to the temperate phage integration site. We need to be able to **select for inactivation of the gene** into which we want the F′ to have integrated, and this means a gene involved in the synthesis of a virulent phage receptor site. Inspection of the map on page 609 shows that there is only one receptor site gene **directly adjacent to a phage integration site**, and that is the Dobbs-*rec* locus, which is right next to *att-Charlie Allnut*. So, we would then screen for Maltese Falcon$^+$ cells (laser-induced red colonies) at 42°C after infection with the virulent Dobbs phage. Surviving red colonies would have to be resistant to the Dobbs phage and have the F′ *Maltese Falcon* integrated into the chromosome. The only single event that could result in this phenotype would be the desired integration into the Dobbs-*rec* locus.

The next step would be to lysogenize with the Charlie Allnut phage. Then, we would induce the phage and make a lysate. We will assume that we can induce the phage with UV. The lysate is then used to transduce a strain that is deleted for the *Maltese Falcon* gene to Maltese Falcon$^+$. If we had a direct selection, such as growth on minimal medium without "Maltese Falcon" supplement, we would select for the rare transducing phage and then take individual colonies and again induce them to verify that we now have a high frequency transducing phage. If we were to conform to the exact wording in the problem as to what is supplied, then we would have to screen transduced cells for colonies that turn red when exposed to a 340-nm frequency laser beam instead of selecting. The principle is the same, except that we would have to look at colonies on many plates instead of examining only a few plates.

3. Here, we would utilize the strain carrying point mutations in each of the genes described on page 608 as a recipient in a cross with the Hfr generated as a first step in the above cloning of the *Maltese Falcon* gene. A true integration of the F′ *Maltese Falcon* into the Dobbs-*rec* gene would generate an Hfr that would donate markers in a fixed direction from that point. Therefore, one easy experiment would be to do a relatively short mating (30 minutes) with the Hfr and the multiply marked recipient and to plate on medium selecting for markers on either side of the presumed integration point. Thus, the *Blue Parrot* and *Seawitch* markers on one side of the integration point and the *Dark Passage* and *Sirocco* markers on the other side should be tested. The single mating mix would be plated on four types of medium, each containing all of the nutrients required for growth of the multiply marked strain except for one (either *Blue Parrot, Seawitch, Dark Passage* or *Sirocco*). The expected result is a high frequency of recombinants for the markers on one side of the integration point or the other, but not both or neither.

A. For the *African Queen:*

1. The principle is the same as described for the *Maltese Falcon,* except that we have to use a different virulent phage. Note that the *African Queen* gene is two genes away from the Captain Queeg-*rec* locus. Thus, we select for resistance to the Captain Queeg phage and examine the surviving colonies for those that do not turn red when exposed to the appropriate laser frequency. The phage-resistant mutants that are African Queen$^-$ have deletions of part or all of this locus.

2. Again, the same principles apply as for the *Maltese Falcon,* except that we use different phage. To isolate an Hfr as a first step, we select strains in which the F factor has integrated into the Captain Queeg-*rec* locus. To do this, we use the deletion strain from part 1 above. We need to select for Captain Queeg phage-resistant colonies under conditions where the F factor cannot replicate. For this problem, this condition is the presence of 5 μg/ml nalidixic acid. The colonies that grow on unsupplemented medium seeded with the Captain Queeg phage and containing 5 μg/ml nalidixic acid are either single-step resistant mutants that are resistant to the phage or Hfr strains with F integrated into the Captain Queeg-*rec* locus. As stated above for the *Maltese Falcon,* perhaps 1 out of 10 of the survivors are the correct Hfr. If we employed a second selection for those that were still resistant to male-specific phage such as R17 or f2, then we would greatly increase our chances of success. In any case, it is a simple matter to screen a set of surviving colonies for whether they are Hfr strains by using the multiply marked recipient provided. Here, we would plate each mating mixture on medium selecting first for **Sirocco$^+$** and

then for **Dark Passage$^+$**, since these are the markers brought in first and last by the desired Hfr.

The next step is the same as for the *Maltese Falcon*. The Hfr is used in a cross with a Rec$^-$ strain carrying a deletion of the *African Queen* locus. Strictly speaking, we cannot select directly for the African Queen$^+$ phenotype, so we have to do a cross in which the donors are killed by either streptomycin or a virulent phage to which the recipient is made resistant (see *Maltese Falcon*). We would then examine colonies to find those that turn red when exposed to the laser. This would require screening many colonies, since the F′ derivative is rare. A better method in this case is to take advantage of the fact that we can select for Sirocco$^+$ strains from a Sirocco$^-$ background using the strains and medium described on page 608. Since the *Sirocco* locus is just before the *African Queen* locus, we would then select for rare F′ *Sirocco* derivatives after a cross of the Hfr into a Rec$^-$ Sirocco$^-$ African Queen$^-$ (deletion) strain. We would then screen the strains carrying the F′ *Sirocco* factor for those that have also become African Queen$^+$ by looking for laser-induced red colony formation. These are the strains carrying an F′ that contains both the *Sirocco* and *African Queen* genes.

Cloning the *African Queen* Gene: The steps are virtually identical to those described for cloning the *Maltese Falcon* gene. We start out with the strain deleted for the *African Queen* gene. We select for integration events into the Dobbs-*rec* gene by plating under conditions where the F′ *African Queen* factor cannot replicate (in the presence of 5 μg/ml nalidixic acid) in the presence of the virulent Dobbs phage and screen for African Queen$^+$ colonies by looking for laser-induced red colonies. Note that if we use the F′ *Sirocco African Queen* factor, then we have to use a Rec$^-$ version of the strain that is deleted for the *African Queen* gene, since otherwise the F′ will reintegrate at the chromosomal *Sirocco* locus using Rec-mediated recombination. The subsequent steps, involving lysogenization with the Charlie Allnut phage, induction, and transduction in two stages, are exactly as described for the *Maltese Falcon* gene.

3. The solution is exactly the same as that described for the *Maltese Falcon* gene.

A. For the *USS Caine*:

1. To isolate a strain carrying a deletion of the *USS Caine* locus, we would infect *Escherichia bogi* with the virulent Ugarte phage, since resistant mutants will have mutations, including large deletions, at the Ugarte-*rec* locus, which is next to the *USS Caine* gene. Phage-resistant mutants are screened for those that are USS Caine$^-$, as determined by their inability to form red colonies when exposed to a 340-nm frequency laser beam.

2. Putting the *USS Caine* gene onto an F′ factor follows the same principles as described for both the *Maltese Falcon* and the *African Queen* genes. First, we generate an Hfr strain with F integrated into the Ugarte-*rec* locus by selecting for resistance to Ugarte phage in the presence of acridine orange, which inhibits the replication of the F factor. If lysates of R17 or f2 are available, we can also select for resistance to male-specific phage. In any case, Ugarte phage-resistant colonies are then tested for whether they are Hfr strains with the origin at the Ugarte-*rec* locus. We use the marked recipients supplied and select first for Sahara$^+$ colonies after a 60-minute or so mating and then, in a second experiment, for Seawitch$^+$. The correct Hfr will yield many recombinants for one marker but not the other.

   The next step also follows along the same lines as for the other two genes. We carry out a mating into a strain deleted for the *USS Caine* gene, and after killing the donor with virulent phage or streptomycin, we then screen for laser-induced red colonies. These are very rare, and it is much better to have a selection. Fortunately, we can select for the *Sahara* gene, which is right next to the *USS Caine* locus. We therefore use the marked recipient, which is Rec$^-$ Sahara$^-$, and isolate our *USS Caine* deletion in that strain. We could even start out with a Rec$^-$ strain and isolate a deletion that covers both the *USS Caine* and *Sahara* genes. The appropriate Rec$^-$ recipient is used to select for Sahara$^+$ colonies, which should result from an F′ factor that carries not only the *Sahara* gene, but also the *USS Caine* gene. The Sahara$^+$ colonies are screened for the laser-induced red colony phenotype to verify the presence of the *USS Caine* gene. The correct F′ factor should transfer the *USS Caine* gene to a second recipient at high frequency.

Cloning the *USS Caine* Gene: The steps are almost identical to those for the *Maltese Falcon* and *African Queen* genes. We start with the strain deleted for the *USS Caine* gene and use the F′ *USS Caine* that we constructed. If we want to use the F′ *USS Caine Sahara* episome, then we need to use a Rec⁻ derivative of the strain deleted for *USS Caine*. In the latter case, we would select for Sahara⁺ colonies in the presence of acridine orange after infection with the Dobbs phage. Alternatively, if we used a Rec⁺ strain and used the simple F′ *USS Caine*, then we would select for colonies in the presence of acridine orange after infection with the Dobbs phage and screen for red colonies in the presence of a 340-nm frequency laser. The subsequent steps involve lysogenization with the Charlie Allnut phage, induction, and transduction, exactly as described for the *Maltese Falcon* gene.

B. To use the Shimada et al. method, we need to isolate a strain with a deletion of the attachment site for a temperate phage and then select for integrations of that phage near the *Maltese Falcon* gene. To generate a deletion of an attachment site, we need to find one that is next to a locus whose inactivation we can select for. Here, we would choose the *att-Charlie Allnut*, since it is adjacent to the Dobbs-*rec* gene, and look at Dobbs-resistant colonies for those that have a lowered frequency of lysogenization with the Charlie Allnut phage, indicating a deletion of the attachment site. We then need to have clear mutants of the Charlie Allnut phage to help select for lysogens. We can mutagenize phage and plate them on a wild-type strain and look for clear mutants that are not clear when plated on a lysogen. These clear mutants are seeded onto a plate as a challenge to cells treated with wild-type Charlie Allnut phage.

For the ***Maltese Falcon*** gene, we also use Señor Ferrari phage, to select for integrations of the Charlie Allnut phage into the Señor Ferrari-*rec* locus that is adjacent to the *Maltese Falcon* gene. We then induce the phage with UV and use the lysate to transduce a strain deleted for the *Maltese Falcon* gene.

For the ***African Queen*** gene, we cannot use a phage selection directly, since it is too far from the gene we want to clone. We have two possibilities. One is to employ a penicillin enrichment for Sirocco⁻ cells and to select for lysogens that are Sirocco⁻. To do this, we need to give the conditions of the penicillin selection (see Unit 4 and the questions pertaining to the papers by Yanofsky and co-workers). We can also isolate a strain with a deletion extending from the *Sirocco* gene to the *Dark Passage* gene, since we can isolate Captain Queeg-resistant colonies and screen them for those that are Sirocco⁻ and Dark Passage⁻. We can then take this strain and select for integrations of the Charlie Allnut phage into the Wilmer-*rec* locus by looking for Wilmer-resistant colonies that are lysogens, surviving the challenge of the clear phage. We now would have the integrated phage adjacent to the *African Queen* gene, and we can induce this phage and use the lysate to transduce a strain with a deletion of the *African Queen* gene.

For the ***USS Caine***, we repeat the procedure for the *Maltese Falcon* gene, except that we use the Ugarte phage and the Ugarte-*rec* locus.

C. For the *Maltese Falcon:*

1. Introducing a charged residue into the interior of the protein can have disastrous effects, since polar groups are excluded from this hydrophobic core.
2. Even though many substitutions of one nonpolar residue for another in the interior of a protein are acceptable, some contacts are so specific that the changes are very restrictive. Here, one would guess that the leucine can make several contacts that the smaller valine cannot.
3. One possibility is that histidine is involved in either the active or binding site of the protein, which often occurs in a crevice, and that tyrosine cannot make the specific contacts involved.
4. Although many exchanges are allowed on the surface of the protein, some residues make contacts with other subunits, and substitutions may not be allowed without losing aggregation.
5. Proline can interrupt α helices and can cause structural distortions and defective proteins even when introduced into some positions on the surface.

C. For the *African Queen:*

1. Proline can interrupt α helices, and its introduction can result in defective proteins due to

these structural deformities even when introduced into some positions on the surface.

2. Some nonpolar residues in the interior make such specific contacts that even a relatively conservative change to another nonpolar residue may not be allowed.
3. Introducing a charged residue into the interior of a protein almost always causes a destabilization, since this is the hydrophobic core.
4. Even though aspartic acid and glutamic acid are similar amino acids, the crevice often contains the active site or the binding site for a ligand, and if a residue such as aspartic acid is making specific contacts in the active site, it may not be substitutable even by glutamic acid.
5. Some residues on the surface of a protein may be involved in subunit interactions and thus may not be replaced easily if they are making specific contacts with another subunit.

C. For the *USS Caine:*

1. Although both isoleucine and phenylalanine are hydrophobic residues, some contacts in the interior of a protein are so specific that this type of change may not be allowed.
2. Introducing a charged residue into the interior of a protein will almost always disrupt the structure, since this is the hydrophobic core of the protein and polar groups are excluded.
3. Introducing proline into an α helix can disrupt the helix and can cause large distortions in the secondary structure. Thus, proline can cause a defective protein when introduced into many positions, even certain surface positions.
4. The crevice of the protein contains the active site or the binding site for a ligand. The tyrosine may be part of the site or may be part of the framework for the site. If the contacts it is making are very specific, then substituting a valine for the tyrosine may destroy the function of the protein.
5. Although many replacements on the surface of a protein are allowed, some residues make contacts with other subunits. Substituting these can destroy aggregation, as may be the case here.

D. In all cases (the *Maltese Falcon*, the *African Queen*, and the *USS Caine*), the procedure would be the same. One needs to start off with a partial diploid carrying two copies of the gene encoding the endorphin-generating protein. After mutagenesis, colonies are screened for those that no longer turn red after exposure to a 340-nm frequency laser beam. The principle here is that the only single mutation that prevents red colony formation would have to be an I^S mutation. This can be proved by showing that it is dominant to I^+ by transferring the episome to another wild-type strain (in cases where the mutation is on the episomal copy) and showing that this strain now also fails to generate red colonies. Alternatively, considering the tools developed, one can bring in yet another copy of the gene encoding the endorphin-generating protein by using the transducing phage constructed in parts A and B and showing that the colonies are not converted to those that turn red after exposure to a laser.

E. Here, we would use the penicillin enrichment technique. After mutagenesis, we would grow the cells for several hours in medium containing all growth requirements and supplements, except for the Letters of Transit supplement, to increase the number of mutants in the population. Under these conditions, Cas$^-$ mutants will stop growing. We then add penicillin and continue to incubate. Since penicillin kills growing cells, it will kill most of the Cas$^+$ cells but not the Cas$^-$ cells. Then, we wash the penicillin out of the medium and add back nutrients, including this time the Letters of Transit supplement. Now, the Cas$^-$ mutants can grow. If we plate the viable cells and test colonies, we can find some Cas$^-$ mutants. Instead of plating, we can increase the percentage of Cas$^-$ mutants among the survivors of the penicillin treatment by going through another cycle of growth without Letters of Transit, followed by penicillin exposure, followed by growth in the presence of Letters of Transit. Several cycles of this enrichment usually results in a significant percentage of the survivors being Cas$^-$.

Scenario 5 / Plantagenet Phage

A. 1. To isolate a complete deletion of the *Agincourt* gene, we take advantage of the fact that mutants resistant to virulent phage often have deletions of the gene encoding the phage receptor site, and some of these deletions extend into and past neighboring genes. Examining the map on page 613, we see that the *Agincourt* gene is adjacent to *Henry V*, which encodes a virulent phage receptor site. The *Troyes* gene is one gene farther away. Deletions into the *Troyes* gene that also eliminate the *Henry V* locus will have to delete the entire *Agincourt* gene. Therefore, we first select Henry V phage-resistant mutants, and then screen them on medium with Troyes supplement and with and without Agincourt supplement, to detect mutants that have at least a partial deletion of the *Agincourt* gene, and also on medium with Agincourt supplement with and without Troyes supplement, to identify Troyes$^-$ strains. The phage-resistant Agincourt$^-$ Troyes$^-$ colonies have complete deletions of the *Agincourt* gene.

2. The first step is to isolate an Hfr strain with F integrated into a locus near the *Agincourt* gene. We do this by introducing the temperature-sensitive F factor into *E. plantagenet*. We then select at 42°C for mutants resistant to Henry V phage. Some fraction of the resistant mutants result from the integration of the F factor into the *Henry V* locus. We can screen a set of resistant mutants for those that are Hfr strains with points of origin at the *Henry V* locus. (If we had male-specific phage lysates available, such as those of R17 or f2, then we could simultaneously select for maintenance of F together with resistance to phage Henry V.) To do this, we would use a Str^r recipient that was defective in a gene near each side of the putative integration point. By extending the same principle we used to find a deletion of the *Agincourt* gene, we can use resistance to the John of Gaunt phage to find Normandy$^-$ mutants, and resistance to the Henry VI phage to find Loire$^-$ mutants. Alternatively, we could use a penicillin selection to enrich for these mutants. The correct Hfr would give a high level of recombinants for one marker (Loire or Normandy) but not the other.

The next step is to use the Hfr in a cross against a Str^r Rec$^-$ derivative of the strain carrying a deletion of both *Troyes* and *Agincourt* that we generated in part 1. The cross would be plated on minimal medium containing Troyes supplement but lacking Agincourt supplement. Any rare Agincourt$^+$ colonies probably have an F′ *Agincourt*, which can be verified by its high frequency of transfer of the Agincourt$^+$ character to an Agincourt$^-$ strain.

3. The key point here is to find an attachment site for a potential transducing temperate phage that is located next to a selectable marker, such as a phage resistance locus. We note that *att-Joan of Arc* is located next to the *Earl of Salisbury* phage receptor gene. Therefore, we start with the strain deleted for the *Agincourt* gene and carrying the temperature-sensitive F′ *Agincourt*, infect with Earl of Salisbury phage, and select colonies at 42°C that are resistant to the phage. These could be tested to see whether they are Hfr strains donating a marker on one side of this locus or the other, but not both (see above). The next step is to lysogenize with the Joan of Arc phage and then to UV-induce the lysate to transduce the *Agincourt* deletion strain to Agincourt$^+$. The rare Agincourt$^+$ colonies are then grown up and UV-induced again, this time yielding, if correct, a high frequency transducing lysate for the *Agincourt* gene.

B. We first need to delete the normal attachment site for The Dauphin temperate phage. Since *att-The Dauphin* is near the *Henry VI* phage receptor locus, we select for mutants resistant to phage Henry VI and screen for those that cannot grow on medium without the Vermeuil supplement. Such strains have a deletion extending from *Henry VI* to *Vermeuil* and must therefore eliminate *att-The Dauphin*. We then take this strain and infect with The Dauphin phage and spread on plates seeded with two different phage: (a) A derivative of The Dauphin phage that lacks immunity and cannot integrate into the chromosome. This will kill all cells that have not become lysogens, or which are resistant. (b) Regent Bedford virulent phage. This selects for mutants resistant to Regent Bedford phage, which have defects or interruptions in the *Regent Bedford* locus next to the *Tours* gene. We can verify that we have a lysogen by UV-inducing surviving colonies and showing that we get a phage lysate for The Dauphin phage. This lysate is used to transduce a Tours$^-$ strain to Tours$^+$. We can construct the Tours$^-$ strain by taking *E. plantagenet* and isolating Regent Bedford phage-resistant mutants and screening for those that are Tours$^-$ (cannot grow on minimal medium without the Tours supplement). The rare Tours$^+$ colonies are then grown up and again UV-induced, and the resulting lysate should show a high frequency of transduction of a Tours$^-$ strain to Tours$^+$.

C. In the recipient that carries the Jean the Good phage at its normal attachment site, markers are inherited normally, since there is phage immunity in the cytoplasm that prevents induction of the incoming phage. That is, the markers closest to the origin will be inherited most frequently, but all markers will be inherited at a characteristic frequency. In the cross involving the recipient that does not carry any prophage, the genetic markers *Tours*$^+$, *Sluys*$^+$, and *Calais*$^+$ will be inherited normally, since they are transferred before the prophage, and in most cases, they will be part of a fragment that broke off before the prophage was transferred. However, the markers entering after the prophage (in this case, *La Rochelle*$^+$ and *Bordeaux*$^+$) are transferred on the same fragment that carries the prophage. Since there is no immunity in the recipient, the prophage is induced, killing the cell. Very few surviving colonies will inherit these late markers. For the *Poitiers*$^+$ gene, the situation is somewhat different. Even though it enters before the prophage, it is so closely linked to the integrated prophage that very few colonies will have received this gene and not the prophage. Therefore, the inheritance will be severely reduced.

D. A mutation in the recipient that deletes the attachment site for the Joan of Arc phage will not affect transduction by P1 at all. However, it will virtually eliminate specialized transduction, since the inheritance of the *Troyes*$^+$ gene depends on the transducing phage integrating into the phage attachment site.

A mutation in the donor strains that prevents DNA digestion will eliminate generalized transduction by P1, which digests DNA into fragments, some of which are incorporated into phage heads. Specialized transducing formation should be unaffected.

A Rec$^-$ mutation will eliminate generalized transduction, since inheritance of markers depends on homologous recombination. Specialized transduction will not be greatly affected, since the phage integration system catalyzes the incorporation of the specialized transducing phage into the chromosome.

Scenario 6 / Globin Wars

The principles of protein structure elucidated by the work of Perutz and his colleagues are straightforward. Polar residues are excluded from the interior of the protein. The three-dimensional structure of a protein draws heavily on the pattern of hydrophobic residues in the interior of the protein. Many substitutions are acceptable on the exterior of the protein unless they interfere with subunit contacts, disrupt the secondary structure of the protein by interrupting helices, as in the case of proline, or affect the substrate or ligand binding site. The latter site is usually in a crevice of the protein. Residues that are important for the active site of the protein can be very sensitive to substitution. In the interior, certain hydrophobic exchanges are acceptable, but others are not. Introducing a polar residue into the interior destabilizes the protein.

A. Considering the principles above, we can examine the variation of residues at specific sites in the galactoglobin molecule. Here, we are asked to explain where the residue at that position is likely to be in the protein based on this pattern of variation.

Position 27: Note that the only residues that appear here are hydrophobic. Therefore, the residue at this position is likely to be in the interior of the protein.

Position 42: Like position 27, the only residues here are hydrophobic, so again we are dealing with an interior site.

Position 58: This position is also on the interior of the protein, and much more conserved, indicating that it makes a crucial contact.

Position 68: The amino acids here are charged or polar, and quite varied. Therefore, the site must be on the exterior.

Position 110: The polar amino acid taurusine is invariant here. This suggests that it is in or near the active site in a crevice of the protein.

B. Looking at the altered galactoglobins from the five patients, we can suggest the following reasons for why each might be defective.

Gb Aldebaran: The change from a polar to a nonpolar amino acid in the crevice might affect the binding of galactin, since the binding site is located in the crevice.

Gb Endor: Normally, changes on the exterior are relatively innocuous. The change from canisine to dracodine might affect subunit-subunit binding, but it is perhaps more likely that the dracodine, in analogy with proline, disrupts the secondary structure of the protein.

Gb Tatooine: Here, the introduction of a charged residue into the interior of the protein will destabilize the structure.

Gb Pit of Sarnak: The change of one polar amino acid to another on the exterior of the protein would not normally be expected to generate a defective protein unless the original amino acid at this position makes an important contact with another subunit.

Gb Death Star: Although sometimes one hydrophobic residue can be substituted for another in the interior, this is not always the case. The sizes of the residues might be significantly different and important contacts might be disrupted, as would be the case here.

Scenario 7 / The Lost Amino Acids of Atlantis

A. 1. If we consider the principles from Pertuz's work discussed on pages 289–290 of the text, we can then look at the first chart and determine whether the positions are likely to be in the interior or on the exterior of the protein. If we examine position 1 in the Hb from each of the creatures (Centaur, Triton, etc.), we see that only three amino acids are found there, Hector, Zeus, and Cassandra. This is in contrast to the residues seen at position 2, where 11 different amino acids are seen. Clearly, position 2 is on the exterior of the protein and position 1 is probably in the interior. Therefore, the residues found at position 1 are nonpolar. Similarly, we can tentatively assign positions 4 and 6 to the exterior and positions 3, 5, and 7 to the interior. Each of the amino acids found in interior positions would be nonpolar and the remaining residues would be polar. At this stage, we can list the amino acids as follows:

Nonpolar	*Polar*
Hector	Apollo
Zeus	Poseidon
Cassandra	Agamemnon
Paris	Hecuba
Calchas	Adonis
Hermes	Patroclus
Achilles	Ajax
Odysseus	Minerva
Aphrodite	Iphigenia
Helen	Priam

2. We can then refer to the chart of altered hemoglobins to verify our assignments. Thus, Hb Sparta involves changing Cassandra to Iphigenia at position 1. We have designated this position as an interior position, and changing a nonpolar to a polar residue would certainly disrupt the protein. With only a few exceptions, all of the amino acid changes confirm our assignments. The three exceptions involve Hb Sybaris, Hb Thrace, and Hb Tartarus. In each case, a substitution at an exterior position causes a defective protein, even though many residues can occupy that position, as seen in the preceding table. However, in all three cases, the new amino acid is Helen. This suggests that Helen = proline, since the introduction of proline can indeed cause a disruption of the structure, even if it involves an exterior position. Thus, we conclude that **Helen** corresponds to proline.

B. 1. By using the same logic as described for part A, we would assign positions 1, 2, 4, and 7 as interior positions, and the amino acids at these positions as nonpolar. Positions 3, 5, and 6 would be exterior positions. The final assignments are as follows:

Nonpolar	*Polar*
Hermes	Ajax
Aphrodite	Minerva
Zeus	Patroclus
Paris	Iphigenia
Helen	Adonis
Cassandra	Priam
Calchas	Hecuba
Hector	Poseidon
Achilles	Apollo
Odysseus	Agamemnon

2. Note that here the same residues are polar and nonpolar as in A above, although the positions are different. However, in looking at the second chart in part B, we see that although we can confirm these assignments, there are three exceptions, all involving changes to Achilles. So, by the same logic as described for A above, we can designate **Achilles** as proline.

Scenario 8 / The Hundred Years Hemoglobin War

The concept of this problem is identical to that in Scenario 7, except that here the amino acids are named after Kings and Queens of England.

1. Looking at the chart showing the seven different positions of hemoglobin in different vertebrates, we can see that position 25 is occupied by either Richard I or John. Thus, we assign this to an interior position for the present and presume that Richard I and John are nonpolar residues. Similarly, positions 56, 108, and 138 should also be interior. The other positions have many residues and are on the exterior. The polar and nonpolar residues are:

Nonpolar	*Polar*
Richard I	Edward I
John	Philippa
Berengaria	William II
William I	Henry III
Henry I	Matilda of Flanders
Isabella of Angoulême	Matilda of Scotland
Eleanor of Aquitaine	Eleanor of Provence
Henry II	Edward III
Edward II	Isabella of France
Stephen	Eleanor of Castille

2. Checking the second chart, where altered hemoglobins are considered, we see that our assignments are logical, but there are exceptions for Hb Gloucester, Hb Lancaster, and Hb March, where in each case a change at an exterior position causes a defective protein. In all three cases, this change is to Berengaria. Thus, we designate **Berengaria** as proline, since as in the previous Scenario 7, we recognize that proline can disrupt the secondary structure even when introduced at certain exterior positions.

Scenario 9 / Mission Impossible: To Decode Deadly Genotypes

Let's solve one of the ciphers, step by step, to demonstrate the procedure for working with cryptograms of this nature. Let's take case **B** on page 626, which is reproduced here as Figure 1. Note that here we are dealing with only two symbols, "+" and "−." What we see is a linear array of "+s" and "−s." The first thing we have to decide is how many strains or symbols stand for each code letter. If only one strain was used, then we would only encode two letters of the alphabet, since each strain can only be either "+" or "−." Similarly, two strains could encode four letters of the alphabet, three strains eight letters, and so on. Clearly, to be able to encode all 26 letters of the alphabet, we need at least five strains, since only 16 letters could be encoded by four strains. Therefore, our first task is to recognize that each group of five symbols equals one letter and to mark off each letter starting from the beginning of the message (as done in Figure 2) and make a frequency chart. This chart is shown in Figure 3.

− + − + + − + − + − + + − − − + − + − + − + − − + − + − + − + + − − − + + − + − − − + + − − − + + − − + − + − + − − + +

+ − + − − + + − − − − + − + − − − + + − − + − + − − + − − + + + − + − − − + + + − + − + − + + − + − + − − + + + − − − +

− + − + − − + + + − + − + − + + + − + − + + − + + + − − + + − + − + − + + + − + + − + − − + − + − − + + + − − + − + − −

+ + − − − − + − + − + + − − − + + + − − − + + − − + − − + − − + − + − + − + − − + + − − − + − + − + + − + − − + − − − +

− + − + − − + + + − + + + − − + − − + + − + + + + + − − + − + − + − − + + + − − + − + − + + + + + + + + − + − − + + +

− + − + + + − + + + − − + − + − − + + + + − + − + − + − − + + − − + + − + + − + − + − + − − + + − + − + − − + − + + − +

− + − − + + + − + − − + + − − − − + − + − + − − − + + − + − + − − + + + − + + − + + − − + − − + − + − + − + − + + − + −

− + + + − − + + + + + + + − − − − + − + − + + − + + + + − + + + + + − − − + + + − − + + + − + − − + − + + − + + + + +

− − + − − − − − − + + + − + − + − − − + + − + + − + − + − + − + − + + + − + − + − − + + − − − − + − + − − − + + + + + +

− − − − + + − − + − + − − + − + − + + − + − − + + + + + + − − − + − + − − − + − + − − + − + − + − − + + + + − + + + + −

+ − − + + + + + + − − − + − + − − + − − + − + + − + − + + + − + + − + − − − + − + − − + − + − + − − + + − + + + − − + −

+ − + + + + + + − + − − + − + − − − + − + + − + − + − + − − + + + − + − − + − − − − + + + − − + − − + + − + − + − + + −

+ + + + − + − − + + + + + + − − − + − + − − + − + − − + + + − − − − + + − + + − + − + − − + − + − + − + + + − + + + − +

+ +

Figure 1

```
− + − + + | − + − + − | + + − − − | + − + − + | − + − − + | − + − + − | + + − − − | + + − + − | − − + + − | − − +
+ − | − + − + − | + − − + + | + − + − − | + + − − − | − + − + − | − − + + − | − + − + − | − + − − + | + + − + − |
− − + + + | − + − + − | + + − + − | + − − + + | + − − − + | − + − + − | − + + + − | + − + − + | + + − + − | + + −
+ + | + − − + + | − + − + − | + + + − + | + − + − − | + − + − − | + + + − − | + − + − − | + + − − − | − + − + − |
+ + − − − | + + + − − | − + + − − | + − − + − | − + − + − | + − + − − | + + − − − | + − + − + | + − + − − | + − −
− + | − + − + − | − + + + − | + + + − − | + − − + + | − + + + + | + − − + − | + − + − − | + + + − − | + − + − + |
+ + + + + | + + − + − | − + + + − | + − + + + | − + + + − | − + − + − | − + + + + | − + − + − | + − − + + | − − +
+ − | + + − + − | + − + − − | + + − + − | + − − + − | + + − + − | + − − + + | + − + − − | + + − − − | − + − + − |
+ − − − + | + − + − + | − − + + + | − + + − + | + − − + − | − + − + − | + − + − + | + − + − − | + + + − − | + + +
+ + | + + − − − | − + − + − | + + − + + | + + − + + | + + + − − | − + + + − | − + + + − | + − − + − | + + − + + |
+ + + − − | + − − − − | − − + + + | − + − + − | − − + + − | + + − + − | + − + − + | − + + + − | + − + − − | + + −
− − | − + − + − | − − + + + | + + + − − | − − + + − | − + − + − | − + − + − | + + − + − | − + + + + | + + − − − |
+ − + − − | − + − + − | − + − + − | + − − + + | + + − + + | + + − + − | − + + + + | + + − − − | + − + − − | + − −
+ − | + + − + − | + + + − + | + − + − − | − + − + − | − + − + − | + − − + + | − + + + − | − + − + − | + + + + + |
+ − + − − | + − + − − | − + − + + | − + − + − | + − − + + | + − + − − | + − − − − | + + + − − | + − − + + | − + −
+ − | + + − + + | + + − + − | − + + + + | + + − − − | + − + − − | + − + − − | + + + − − | − − + + − | + + − + − |
+ − − + − | + − + − + | + + − + + | + − + + + |
```

Figure 2

Symbol	Occurrences	Rank	Symbol	Occurrences	Rank
+ + + + +	3	14	− + + + −	8	8
+ + + + +	0		+ + − − −	12	4
+ + + − +	2	16	+ − + − −	19	2
+ + − + +	7	10	+ − − + −	7	9
+ − + + +	2	17	− + − + −	27	1
− + + + +	5	12	+ − − − +	3	15
+ + + − −	10	5	− + − − +	2	19
+ + − + −	14	3	− − + − +	0	
+ − + + −	0		− − − + +	0	
− − + + −	7	11	− + + − −	1	22
+ + − − +	0		− − − − +	0	
− + + − +	1	21	− − − + −	0	
+ − + − +	8	7	− − + − −	0	
+ − − + +	9	6	− + − − −	0	
− + − + +	2	18	+ − − − −	2	20
− − + + +	4	13	− − − − −	0	

Figure 3 Frequency chart for code B

18—1—4—7—19—1—4—3—11—11—1—6—2—4—1—11—1—19—3—13—1—3—6—15—1—8—

7—3—10—6—1—16—2—2—5—2—4—1—4—5—22—9—1—2—4—7—2—15—1—8—5—6—12—

9—2—5—7—14—3—8—17—8—1—12—1—6—11—3—2—3—9—3—6—2—4—1—15—7—13—

21—9—1—7—2—5—14—4—1—10—10—5—8—8—9—10—5—20—13—1—11—3—7—8—2—4—

1—13—5—11—1—1—3—12—4—2—1—1—6—10—3—12—4—2—9—3—16—2—1—1—6—8—1—

14—2—2—18—1—6—2—20—5—6—1—10—3—12—4—2—2—5—11—3—9—7—10—17

Figure 4 Transposed message

At this point, we can simplify our task by assigning a number to each group of five symbols. This is solely to avoid the strain of constantly recognizing each group of five symbols and is purely optional. In addition, if we rank the groups according to frequency, then the number also allows us to see the relative frequency right on the linear message itself. This renumbering, giving each symbol type a number corresponding to its frequency rank, transposes the message to that shown in Figure 4. We can now proceed to solve the message in a series of steps.

1. Referring to the frequency chart and recalling from page 625 that the frequency of single letters in the English language is E > T > A > O > N > I > R > S > H, we note that No. 1 (– + – + –) appears 27 times, No. 2 (+ – + – –) 19 times, and No. 3 (+ + – + –) 14 times. After that the frequencies drop off considerably. Let us assign No. 1 to E, and keep in mind that No. 2 is probably T, simply as a trial. This may not be correct, but it is often a good start, particularly in the case of E. This gives us the situation shown in Figure 5.

E E E E E E E
18—1—4—7—19—1—4—3—11—11—1—6—2—4—1—11—1—19—3—13—1—3—6—15—1—8—

E E E E
7—3—10—6—1—16—2—2—5—2—4—1—4—5—22—9—1—2—4—7—2—15—1—8—5—6—12—

E E E
9—2—5—7—14—3—8—17—8—1—12—1—6—11—3—2—3—9—3—6—2—4—1—15—7—13—

E E E
21—9—1—7—2—5—14—4—1—10—10—5—8—8—9—10—5—20—13—1—11—3—7—8—2—4—

E E E E E E E E
1—13—5—11—1—1—3—12—4—2—1—1—6—10—3—12—4—2—9—3—16—2—1—1—6—8—1—

E E
14—2—2—18—1—6—2—20—5—6—1—10—3—12—4—2—2—5—11—3—9—7—10—17

Figure 5

```
   E H      E H         E   T H E    E         E        E
18—1—4—7—19—1—4—3—11—11—1—6—2—4—1—11—1—19—3—13—1—3—6—15—1—8—

         E    T T   T H E H        E T H   T    E
7—3—10—6—1—16—2—2—5—2—4—1—4—5—22—9—1—2—4—7—2—15—1—8—5—6—12—

  T                 E    E        T         T H E
9—2—5—7—14—3—8—17—8—1—12—1—6—11—3—2—3—9—3—6—2—4—1—15—7—13—

     E   T      H E                           E          T H
21—9—1—7—2—5—14—4—1—10—10—5—8—8—9—10—5—20—13—1—11—3—7—8—2—4—

E         E E      H T E E           H T         T E E     E
1—13—5—11—1—1—3—12—4—2—1—1—6—10—3—12—4—2—9—3—16—2—1—1—6—8—1—

   T T    E   T        E         H T T
14—2—2—18—1—6—2—20—5—6—1—10—3—12—4—2—2—5—11—3—9—7—10—17
```

Figure 6

2. Let's see if there are any good candidates for the word "the." We note the frequencies of sequences of X-Y-E, with X and Y being the number preceding E, and see that 2-4-E appears four times. Since No. 2 is likely to be T based on single letter frequencies (see above), let's now assign No. 2 to T and 4 to H. This gives us Figure 6.

3. Examining Figure 6 allows us to guess the identity of several letters. In two cases, we see the combination "H-T." In one case, it is part of 12-H-T-E-E-, and in another case it is part of 12-H-T-T-. In both cases, the T after the H is unlikely to or cannot begin a word, meaning that the letter before -H-T- is almost certainly a "G." Note that it is No. 12 in both cases, and the letter before it should be "I" (No. 3 in both cases). In the first case, this gives -E-E-I-G-H-T-E-E-6-, clearly pointing toward the word "EIGHTEEN" and allowing us to assign No. 6 to N. Figure 7 shows the advance in decoding achieved with the assignments of I, G, and N.

```
   E H      E H I       E N T H E    E    I    E I N    E
18—1—4—7—19—1—4—3—11—11—1—6—2—4—1—11—1—19—3—13—1—3—6—15—1—8—

  I    N E    T T   T H E H        E T H   T    E     N G
7—3—10—6—1—16—2—2—5—2—4—1—4—5—22—9—1—2—4—7—2—15—1—8—5—6—12—

  T        I        E G  E N    I T I   I N T H E
9—2—5—7—14—3—8—17—8—1—12—1—6—11—3—2—3—9—3—6—2—4—1—15—7—13—

     E   T      H E                           E    I     T H
21—9—1—7—2—5—14—4—1—10—10—5—8—8—9—10—5—20—13—1—11—3—7—8—2—4—

E         E E I G  H T E E N    I G  H T   I    T E E N   E
1—13—5—11—1—1—3—12—4—2—1—1—6—10—3—12—4—2—9—3—16—2—1—1—6—8—1—

   T T    E N T      N E    I G  H T T      I
14—2—2—18—1—6—2—20—5—6—1—10—3—12—4—2—2—5—11—3—9—7—10—17
```

Figure 7

E H E H I E N T H E E I E I N E
18—1—4—7—19—1—4—3—11—11—1—6—2—4—1—11—1—19—3—13—1—3—6—15—1—8—

I R N E X T T T H E H S E T H T E N G
7—3—10—6—1—16—2—2—5—2—4—1—4—5—22—9—1—2—4—7—2—15—1—8—5—6—12—

S T I E G E N I T I S I N T H E
9—2—5—7—14—3—8—17—8—1—12—1—6—11—3—2—3—9—3—6—2—4—1—15—7—13—

S E T H E R R S R E I T H
21—9—1—7—2—5—14—4—1—10—10—5—8—8—9—10—5—20—13—1—11—3—7—8—2—4—

E E E I G H T E E N R I G H T S I X T E E N E
1—13—5—11—1—1—3—12—4—2—1—1—6—10—3—12—4—2—9—3—16—2—1—1—6—8—1—

T T E N T N E R I G H T T I S R
14—2—2—18—1—6—2—20—5—6—1—10—3—12—4—2—2—5—11—3—9—7—10—17

Figure 8

4. Figure 7 offers several new easy assignments. The configuration -I-T-I-9-I-N-T-H-E- must be "IT IS IN THE," which allows us to allocate S to No. 9. This makes it easy to see that the stretch "E-I-G-H-T-E-E-N-10-I-G-H-T-S-I-16-T-E-E-N" reveals No. 16 to be "X," to give "SIXTEEN," and let's guess that No. 10 is "R," to give "EIGHTEEN RIGHT SIXTEEN." We could even speculate that "18-E-14-T" is "LEFT," but let's first fill in the S, R, and X assignments to generate Figure 8.

5. From Figure 8, we can make several informed guesses. The word "next" in the sequence, "N-E-X-T-T-5-T-H-E-H-5-22-S-E-T-H-7-T-" suggests that No. 5 = the letter O, and leaves us with "NEXT TO THE HO22SE TH7T," which lets us assign No. 22 to "U" and No. 7 to "A." These new assignments generate Figure 9.

6. This is virtually enough information to solve the code. Starting near the beginning, the double let-

E H A E H I E N T H E E I E I N E
18—1—4—7—19—1—4—3—11—11—1—6—2—4—1—11—1—19—3—13—1—3—6—15—1—8—

A I R N E X T T O T H E H O U S E T H A T E O N G
7—3—10—6—1—16—2—2—5—2—4—1—4—5—22—9—1—2—4—7—2—15—1—8—5—6—12—

S T O A I E G E N I T I S I N T H E A
9—2—5—7—14—3—8—17—8—1—12—1—6—11—3—2—3—9—3—6—2—4—1—15—7—13—

S E A T O H E R R O S R O E I A T H
21—9—1—7—2—5—14—4—1—10—10—5—8—8—9—10—5—20—13—1—11—3—7—8—2—4—

E O E E I G H T E E N R I G H T S I X T E E N E
1—13—5—11—1—1—3—12—4—2—1—1—6—10—3—12—4—2—9—3—16—2—1—1—6—8—1—

T T E N T O N E R I G H T T O I S A R
14—2—2—18—1—6—2—20—5—6—1—10—3—12—4—2—2—5—11—3—9—7—10—17

Figure 9

W E H A V E H I D D E N T H E D E V I C E I N B E L
18—1—4—7—19—1—4—3—11—11—1—6—2—4—1—11—1—19—3—13—1—3—6—15—1—8—

A I R N E X T T O T H E H O U S E T H A T B E L O N G
7—3—10—6—1—16—2—2—5—2—4—1—4—5—22—9—1—2—4—7—2—15—1—8—5—6—12—

S T O A F I L M L E G E N D I T I S I N T H E B A C
9—2—5—7—14—3—8—17—8—1—12—1—6—11—3—2—3—9—3—6—2—4—1—15—7—13—

K S E A T O F H E R R O L L S R O Y C E D I A L T H
21—9—1—7—2—5—14—4—1—10—10—5—8—8—9—10—5—20—13—1—11—3—7—8—2—4—

E C O D E E I G H T E E N R I G H T S I X T E E N L E
1—13—5—11—1—1—3—12—4—2—1—1—6—10—3—12—4—2—9—3—16—2—1—1—6—8—1—

F T T W E N T Y O N E R I G H T T O D I S A R M
14—2—2—18—1—6—2—20—5—6—1—10—3—12—4—2—2—5—11—3—9—7—10—17

Figure 10

ter -11-11 should be "D" to give "HIDDEN." Later, "NEXT TO THE HOUSE THAT 15-E-8-ONGS TO" suggests that 15 = B and 8 = L, to give "BELONGS." We can then see that 14 = F, to give "LEFT" (as we guessed before), 18 = W, and 20 = Y, to give "TWENTY-ONE," and 13 = C, to give "CODE" and "ROLLSROYCE." The final message, which is also shown in Figure 10, is:

We have hidden the device in Bel Air next to the house that belongs to a film legend. It is in the back seat of her Rolls Royce. Dial the code eighteen right sixteen left twenty one right to disarm.

The messages encoded by the other ciphers on pages 626–629 are given below.

A. Each strain has six possible phenotypes. Therefore, two strains can code for 36 letters. This would be analogous to a DNA made up of six different bases. Each position can be occupied by one of six bases. Thus, it is a doublet code. For example, the first letter is encoded by +S CR, the second letter by -S -R, etc. The message, when decoded, reads:

You can find the bomb in Pasadena at the Rose Bowl buried under the goal post at the north end of the field. Disconnect all three electric cables.

C. Each strain has four possible genotypes. Thus, since one strain could encode 4 letters and two could encode 16, then it takes three strains to be able to encode at least 26 letters, since three strains can encode 64 letters. Here, we have a triplet code. The message is:

Proceed West on Sunset Boulevard until you reach the Pacific Coast Highway. The bomb is buried on the beach behind Gladstones fish restaurant, ten meters West and two meters North of the back exit.

D. Each strain has six possible genotypes, so, like code A, it takes only two strains to encode one of 36 possible letters. Thus, we have a doublet code. For example, the first letter is encoded by SN RS. The message reads:

The bomb is hidden in Play Island near the Matterhorn. It is disguised as a spare bobsled and is scheduled to explode three weeks from yesterday evening. It can be defused with the combination three right twenty-one left seventeen right.

E. Each strain can have one of four genotypes. Therefore, we are dealing with a triplet code, where each letter requires three strains. The first letter is encoded by ++ −+ ++. The message reads:

They have stored the device in the La Brea Tar Pits. It is disguised as a Tyrannosaurus Rex. Pull out the three top front teeth to inactivate the explosives.

F. Each strain has 27 possible genotypes. Thus, a single strain can encode one letter. The message reads:

Go to the Santa Monica pier and face the skeeball concession. Look to the right and walk fifteen steps. On the left side there will be a garage with a red sign that says do not enter. The bomb is inside the garage and timed to go off in forty-eight hours. It can be inactivated by dialing the number two hundred thirty-three and then pressing the red button.

G. Here, we have four possible genotypes for each strain. Thus, it takes three strains to encode each letter, a triplet code. The message reads:

The weapon is located in Malibu off Route One at the John Paul Getty Museum under the marble base of the statue of Hercules. Cut the green wire.

Scenario 10 / Jurassic Engineering

The very first thing to be decided in this scenario is the nature of the code. Since there are 34 amino acids and at least one nonsense codon, we need a code that will allow us to specify at least 35 words. We note with satisfaction that a nonoverlapping doublet code will be able to have 36 different codons, since at each of the two positions there are six possible letters. Thus, we start with the assumption that each code word has two letters.

Let's solve part **B** on page 633 in the text step by step to demonstrate the procedure for decoding the toxin gene message. The key point in this problem is that the arrows in level 1 point to changes made by hydroxylamine when the mutagenized viruses are passaged through strain B before plating, whereas in level 2 the arrows point to changes made by hydroxylamine when the mutagenized viruses are passaged through strain K before plating on B. This means that all G·C→A·T transitions will appear in either level 1 and 2 or just in level 1. Any codons that have a G or a C in them will generate a change in level 1 and possibly in level 2 unless that change generates a nonsense codon. Since we are recovering mutants, it is important that the wild-type phenotype is expressed before the first round of replication. Therefore, any Cs in the wild type that are to be changed must be on the antisense strand, so that the change from C to T does not affect the message made off of the sense strand before the first round of replication. Noting that the antisense strand has the same sequence as the message, we can see that codons with a C undergoing the change will generate changes that appear not only in level 1, but also in level 2. Now we can begin to decode the toxin gene message. (For simplicity, let's assume that like the DNA, the message also contains the same six base pairs: T, C, A, G, X, Y.)

1. As a first step, we can pinpoint the codons that are CC, GG, CG, and GC. The CC codon will be the only one that gives two changes in both levels 1 and 2, based on the logic above. From the chart below, we can see that CC must encode TYR, so we make our first codon assignment. By the same token, GG must encode NOR. CG and GC must then encode TRP and VEL, although we do not know at this stage which is which. At this point, we should initiate two things to aid solving the

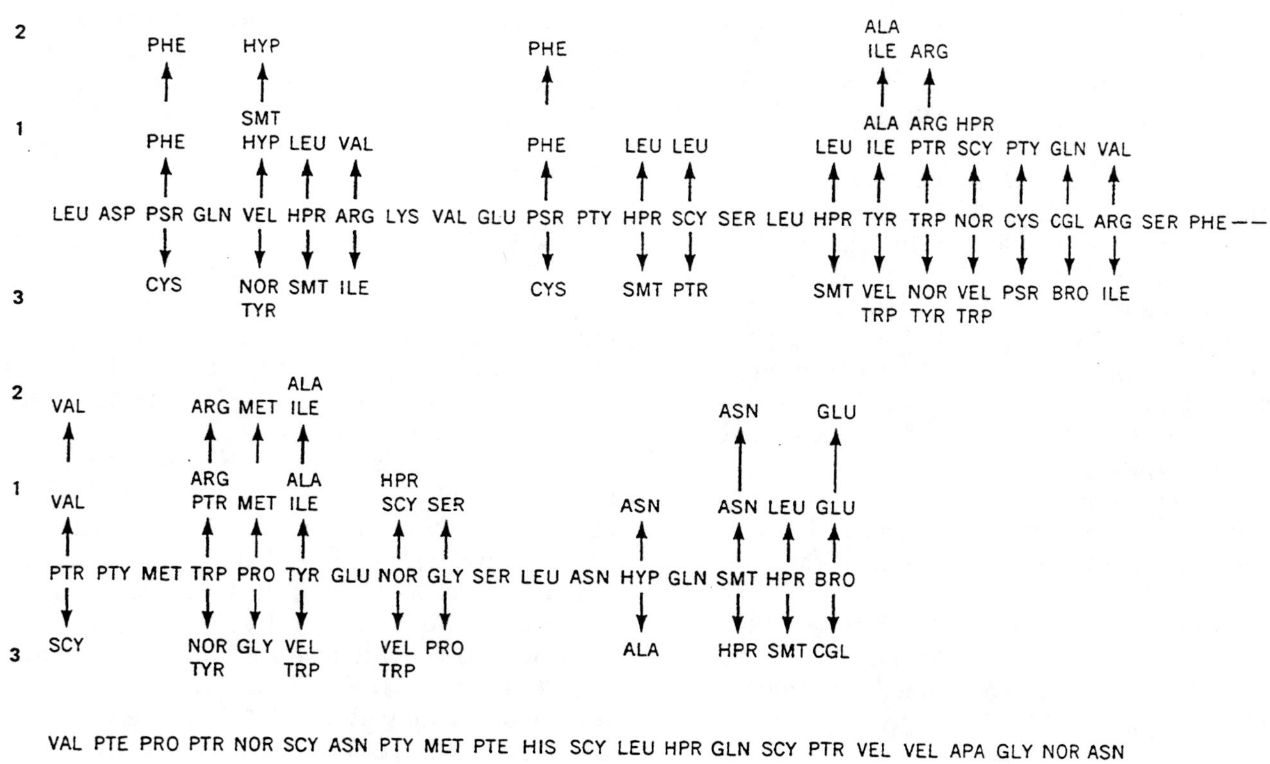

code. First, we should make an emerging codon table (see page 45) to record each new assignment. Second, we should use each new codon assignment to start filling in bases in both the wild-type sequence and in the frameshifted sequence at the bottom of the chart. We can continually use the frameshifted sequence to provide valuable information. Thus, by examining codons 18, 19, and 20, we see that the TYR TRP NOR sequence generates either CC GC GG or CC CG GG, depending on whether TRP is GC or CG. However, the frameshifted sequence, generated by removing the first base of the sequence, will then give as codons 18 and 19 either CG CG or CC GG. Thus, the amino acids at positions 18 and 19 in the sequence will be either VEL VEL or TRP TRP if the sequence is CG CG or will be NOR TYR if the sequence is CC GG. We note that the frameshifted sequence at that position is indeed VEL VEL, which verifies the first possibility and shows that GC specifies TRP and CG specifies VEL. Therefore, at this stage, we already have four code words. We enter them in our code table and also add these new sequences to the wild-type and frameshifted sequences.

2. Now we can look at the mutations arising from our known codons. CG will give rise to TG and CA, with the former appearing in levels 1 and 2 and the latter only in level 1. We can see that at codon 5, VEL gives HYP in both levels, which must be TG, and SMT only in level 1, which must be CA. Likewise, GC will give rise to AC in level 1 and GT in both levels 1 and 2. At codon 19, TRP gives rise to PTR in level 1 and ARG in both levels. Thus, PTR must be AC and ARG must be GT.

   We can then look at the products of mutations at each of the new codons. TG is mutated in level 1 to yield TA. Thus, at codon 38, we see that HYP goes to ASN, so ASN must be TA. CA goes to TA in levels 1 and 2, and we note with satisfaction that at position 40 SMT goes to ASN also. In the same manner, AC and GT should both give rise to AT in levels 1 and 2 or just in level 1. It can be seen that PTR at codon 26 and ARG at codon 7 both give rise to VAL, which must be AT.

   For the remaining codons with only A, T, C, G in them, we can still gain information using only levels 1 and 2. Note that CC goes to CT and TC, which, from codon 18, must be ALA and ILE, although we do not know which is which yet. Both would yield TT in levels 1 and 2, but there is no ALA or ILE in the wild-type sequence. However, GG goes to AG and GA, which, from codon 20, must be HPR and SCY, again not knowing which is which. In this case, both would go to AA in level 1, which, from codon 6, is LEU. Therefore, AA = LEU.

3. Even before we look at level 3, we can deduce other codons by filling in the known sequences in both the wild-type and frameshifted sequences. Thus, by filling in the codons for amino acids 4–7 in both sequences, we can deduce that HPR has G as the first letter of the codon and SCY has G as the second letter of the codon. Since one of these is AG and the other GA (from the last paragraph in step 2), then we can assign GA to HPR and AG to SCY.

4. Now we can look at level 3, both to confirm our assignments and to find CT and TC, two of the remaining three codons in the A, C, T, G series not yet found. Both ALA and ILE, which we have shown are either CT or TC, are generated by the transversion in level 3. CT is generated from GT, and TC is generated from TG. GT is ARG, and we note that at position 7, ARG is converted to ILE in level 3. Therefore, CT = ILE. Thus, ALA must be TC.

5. The next step is to assign the codons containing X and Y. Here we are given a clue, namely, that a mutagen that changed A→X generated a LEU→SER change. Since we have established that AA = LEU, then either AX or XA must equal SER. However, our emerging sequence of the wild-type message made by using both the frameshifted and wild-type sequences has already shown that the first letter of SER must be A (see the SER at position 15 in the wild type and note that the second letter of HPR from the 14th residue of the frameshifted sequence becomes the first letter of SER). Therefore, AX must code for SER. In addition, by plugging the AX for SER into the frameshifted sequence, we can immediately establish that XA = GLN and that X is the first letter of a nonsense codon (the reason for the truncated frameshifted sequence).

   We can then note that GLY is converted to SER in level 1, and thus GX = GLY. In level 3, GLY is converted to PRO, and thus CX = PRO. Likewise, CGL is converted to GLN in level 1 (residue 22), and thus, since XA = GLN, then XG = CGL. PRO is converted to MET in levels 1 and 2, and thus TX = MET. CGL is converted to BRO in level 3, which establishes that XC = BRO. Finally, BRO, at residue 42, is converted to GLU in levels 1 and 2, which means that GLU = XT.

6. By filling in the sequences for the wild-type and frameshifted messages, we can establish several things about the remaining codons. Using a hyphen to indicate an unknown base, we can say that ASP = T-, PSR = -C, LYS = A-, PTY = -A, APA = G-, CYS = -G, PHE = --, one nonsense = X-, and the second nonsense and THR and JUR are --. Since all of the codons containing at least one A, C, T, or G are assigned except for TT and those with a Y, it is clear that we can assign TY or TT = ASP, YC = PSR, AY = LYS, YA = PTY, GY = APA, and YG = CYS.

   By adding these to the sequences, we can derive that TY = PTE and CY = HIS, and from the fact that PSR is converted to PHE in levels 1 and 2, we can say that YT = PHE. By looking at the beginning of the frameshifted sequence, we can now also attribute TT to ASP.

7. This now completes the assignment of each codon except for the codons XX, XY, YX, and YY. At least one of the two codons XY or XX must be nonsense, and the other three must encode THR, JUR, and probably another nonsense signal. With the data given, we cannot go any further. We would need more data from the mutagen that changed A→X to allow us to determine whether the AY and YA codons were converted to either THR or JUR, and if not, then we would assume that they were converted to nonsense. The final code, as well as the mRNA sequence are given below.

TT = ASP	CT = ILE	AT = VAL	GT = ARG
TC = ALA	CC = TYR	AC = PTR	GC = TRP
TA = ASN	CA = SMT	AA = LEU	GA = HPR
TG = HYP	CG = VEL	AG = SCY	GG = NOR
TX = MET	TY = PTE	XT = GLU	YT = PHE
CX = PRO	CY = HIS	XC = BRO	YC = PSR
AX = SER	GY = APA	XG = CGL	YA = PTY
GX = GLY	AY = LYS	XA = GLN	YG = CYS

XX, XY, YX, YY, unassigned; nonsense, THR, JUR, unassigned.

The DNA sequence of the antisense strand, which is identical to the mRNA sequence, is:

A-A-T-T-Y-C-X-A-C-G-G-A-G-T-A-Y-A-T-X-T-Y-C-Y-A-G-A-A-G-A-X-A-A-G-A-C-C-G-C-G-G-Y-G-X-G-G-T-A-X-Y-T-A-C-Y-A-T-X-G-C-C-X-C-C-X-T-G-G-G-X-A-X-A-A-T-A-T-G-X-A-C-A-G-A-X-C

B Genetic Code

1st letter \ 2nd letter	T	C	A	G	X	Y
T	ASP	ALA	ASN	HYP	MET	PTE
C	ILE	TYR	SMT	VEL	PRO	HIS
A	VAL	PTR	LEU	SCY	SER	LYS
G	ARG	TRP	HPR	NOR	GLY	APA
X	GLU	BRO	GLN	CGL	STOP	STOP
Y	PHE	PSR	PTY	CYS	STOP	STOP

The genetic codes for parts **A** and **C–G** of Scenario 10 are given on the following pages.

A Genetic Code

1st letter \ 2nd letter	T	C	A	G	X	Y
T	PRO	HYP	ASP	LEU	GLN	CYS
C	GLU	STOP	TRP	PTE	SCY	CGL
A	ILE	ALA	LYS	HPR	MET	SER
G	APA	BRO	ASN	GLY	NOR	HIS
X	PTY	THR	PSR	VEL	PTR	SMT
Y	JUR	STOP	ARG	TYR	PHE	VAL

C Genetic Code

1st letter \ 2nd letter	T	C	A	G	X	Y
T	APA	LYS	ASN	HIS	TRP	STOP
C	CGL	HPR	ASP	VAL	THR	CYS
A	TYR	NOR	SER	ILE	GLY	SMT
G	ARG	PSR	PTY	GLU	HYP	PRO
X	PHE	LEU	MET	BRO	VEL	PTE
Y	STOP	GLN	STOP	SCY	ALA	PTR

D Genetic Code

1st letter \ 2nd letter	T	C	A	G	X	Y
T	SCY	SMT	GLN	LEU	APA	ALA
C	GLY	PRO	ILE	TRP	PTE	LYS
A	BRO	PHE	HPR	PSR	HYP	NOR
G	VEL	ASP	VAL	PTY	THR	MET
X	PTR	GLU	CGL	ASN	HIS	ARG
Y	STOP	TYR	SER	STOP	CYS	JUR

E Genetic Code

1st letter \ 2nd letter	T	C	A	G	X	Y
T	JUR	SER	PTY	MET	PRO	GLY
C	HIS	GLU	ILE	LEU	VAL	STOP
A	TYR	CYS	TRP	PTE	PTR	ARG
G	SCY	ASN	HYP	NOR	GLN	THR
X	CGL	HPR	PHE	SMT	ALA	VEL
Y	PSR	ASP	LYS	APA	BRO	STOP

F Genetic Code

1st letter \ 2nd letter	T	C	A	G	X	Y
T	PSR	HPR	ILE	JUR	PTE	TYR
C	PTR	ARG	HYP	SER	GLN	CYS
A	SCY	PTY	TRP	ASN	THR	HIS
G	VAL	PRO	LEU	GLU	NOR	CGL
X	APA	ASP	LYS	VEL	PHE	BRO
Y	STOP	STOP	ALA	SMT	MET	GLY

G Genetic Code

1st letter \ 2nd letter	T	C	A	G	X	Y
T	THR	CYS	ILE	PTE	HPR	STOP
C	ASN	TRP	LYS	BRO	PSR	SCY
A	LEU	SMT	ARG	PRO	ALA	PHE
G	PTR	GLY	GLU	JUR	STOP	PTY
X	HIS	CGL	GLN	TYR	ASP	APA
Y	HYP	NOR	VAL	VEL	SER	MET

Scenario 11 / America's Most Wanted: Against All Odds

A. If Hong Kong Maggie bets \$2000 each time and the wheel spins 180 times, then her investment is 180 x \$2000 = \$360,000. If she wins 16 times at 35:1, then she wins 16 x 35 x \$2000 = \$1,120,000. Subtracting her investment, her net profit is **\$760,000.**

The probability that the 35:1 slots come up 16 times in 180 spins can be calculated by first finding m. Here, we note that although there are 120 slots on the wheel, 3 of these pay 35:1. If we place these three slots together, then we have 1/40 of the wheel. In other words, we can envision the wheel as having 40 equivalent slots, one of which pays 35:1. Therefore, m = 180/40 = **4.5**. $P_{16} = m^{16}e^{-m}/16!$ = **1.5 x 10^{-5}**. If we wanted to know the expected number of times that any of the 40 slots would come up as many as 16 times in 180 spins, then we would multiply this by 40, but here we are interested in the 35:1 slots, so we do not multiply the probability by 40. (To be more exact, we should sum the terms for the probability of 16, 17, etc., although these extra terms will not add very much to the total.)

B. If Maggie bets \$2000 for 100 spins, then she invests \$200,000. If she wins 10 times at 50:1, then she wins 10 x 50 x \$2000 = \$1,000,000. Subtracting her investment, her net profit is **\$800,000.**

To find m for the distribution, we note that since there are 2 slots that pay 50:1, we can consider the wheel as being divided into 60 equivalent slots. Therefore, m = 100/60 = **1.667**. $P_{10} = m^{10}e^{-m}/10!$ = **8.62 x 10^{-6}**. Once again, this is the probability that the specific 50:1 slots will appear 10 times out of 100 spins.

C. If Maggie bets \$2000 on each of two cards 150 times, then she invests \$600,000. If she wins 17 times at 50:1, then she wins 17 x 50 x \$2000 = \$1,700,000, for a net profit of **\$1,100,000.**

To calculate m for the distribution, we note that there are 52 cards and that by betting on 2 cards each time, the two cards represent 1/26 of the total cards. Thus, m = 150/26 = **5.77**. $P_{17} = m^{17}e^{-m}/17!$ = **7.6 x 10^{-5}**. This is the probability of her winning 17 of 150 draws by betting on 2 cards each time.

D. If Geek Cassidy bets \$1000 for each of 200 times, then he invests \$200,000. If he wins 25 times at 36:1, then he wins 25 x 36 x \$500 = \$450,000. Subtracting his investment, his net profit is **\$250,000.**

For this distribution, we note that there are 18 intervals equivalent to the two intervals bet on by Cassidy. So, m = 200/18 = **11.1**. $P_{25} = m^{25}e^{-m}/25!$ = **1.3 x 10^{-4}**, which is the probability that one of the two numbers will come up 25 times out of 200 spins.

E. If Geek Cassidy bets \$1000 for each of 200 spins, then his investment is \$200,000. If he wins 34 times at 36:1, then he wins 34 x 36 x \$333.33 = \$407,996. Subtracting his investment, his net profit is **\$207,996.**

Because he bets 3 slots out of 38, then these represent 3/38 of the wheel. We calculate m by dividing 200 spins by 38/3 slots (the number of slots equivalent to the 3 slots he bet on.) Thus, m = 200/12.67 = **15.79**. $P_{34} = m^{34}e^{-m}/34!$ = **2.6 x 10^{-5}**. This is the probability that one of the three numbers bet will come up 34 times out of 200 spins.

Scenario 12 / Hitchhiker's Guide to the Galaxy: Follow That Sequence

A. 1. (*a*) The following sense codons can go to the nonsense codon **QSR**:

TSR, RSR, SSR, QTR, QQR, QRR, QST, QSS, QSQ.

Each of the six base changes results in the following sense codons changing to the nonsense codon QSR, with the number indicating the codon, the first codon in the sequence being numbered "1."

Q•T→S•R = QQR (1, 42, 54, 74, 92, 112, 134), QST (45, 97)
Q•T→R•S = QTR (18, 31, 35, 47, 59, 72, 87), QSQ (106)
Q•T→T•Q = TSR (51, 55, 133)
R•S→Q•T = RSR (38, 80, 101)
R•S→T•Q = SSR (10, 34, 58, 66, 69, 71, 85, 91, 94, 96, 105, 115, 118, 120, 129)
R•S→S•R = QRR (130), QSS (63)

These can be arranged as shown in Figure 1, which diagrams the position of each site.

(*b*) Since the mutagen makes all possible base changes at Q•T base pairs, then only the sites represented by the QQR, QST, QTR, QSQ, and TSR codons will be hit, as seen in Figure 2.

(*c*) The expected results for spontaneous base substitutions would have each of the possible sites at similar frequencies except for hotspots at the methylated bases. To find the methylated bases, we need first to look at the sequence to see whether any QRT sequences overlap with a wild-type codon in such a way that the T will mutate to create the **QSR** nonsense codon. This will only involve the QRTSR sequence, reading the sequence in the 5′→3′ direction, where the TSR is a codon. In other words, we are looking for a -QR TSR. Inspecting each of the three TSR sequences, we note that only one, at position **55**, is preceded by a QR. The T of the TSR will mutate frequently and be a hotspot. So, we will find ambers at this site much more frequently than at other sites.

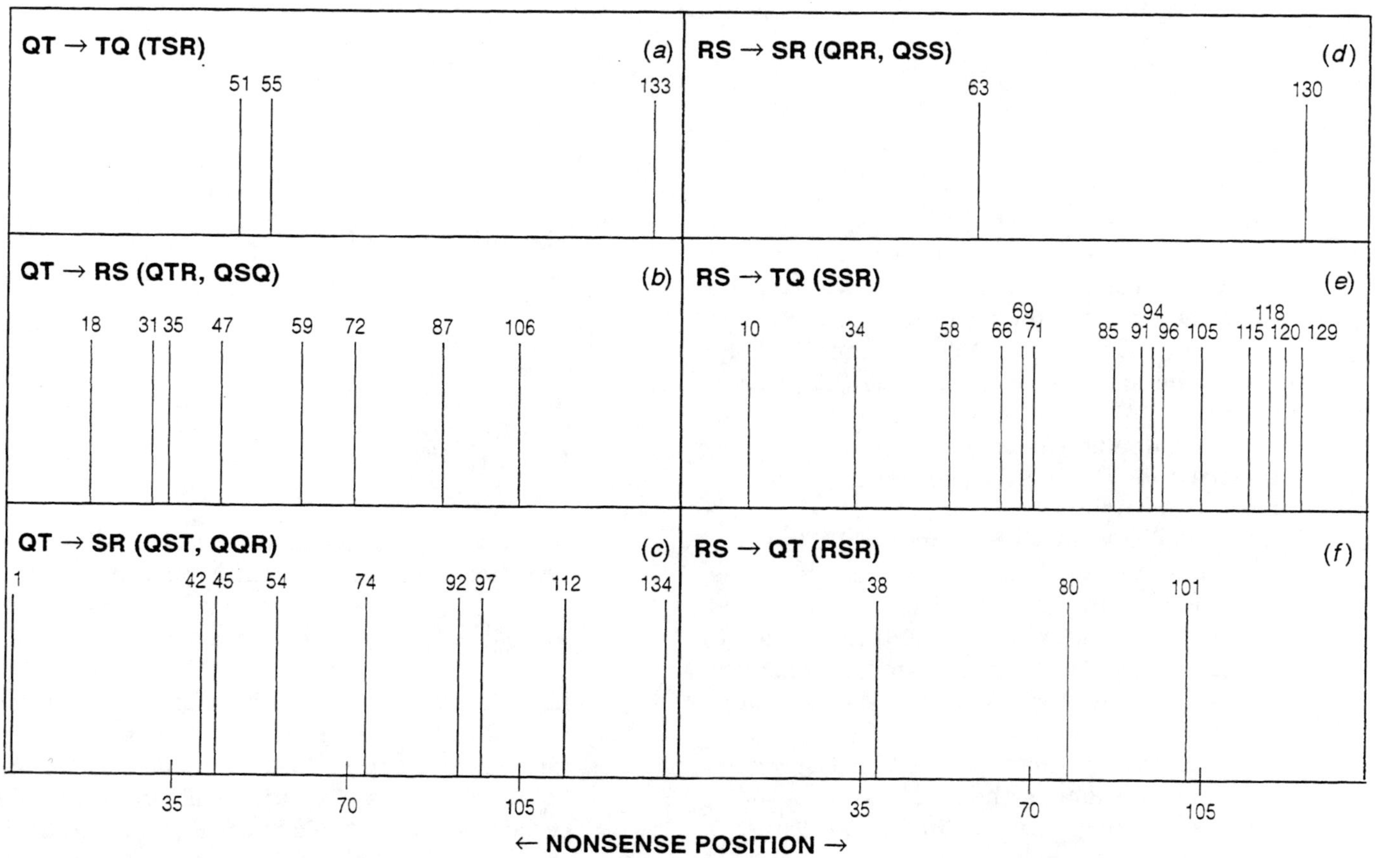

Figure 1

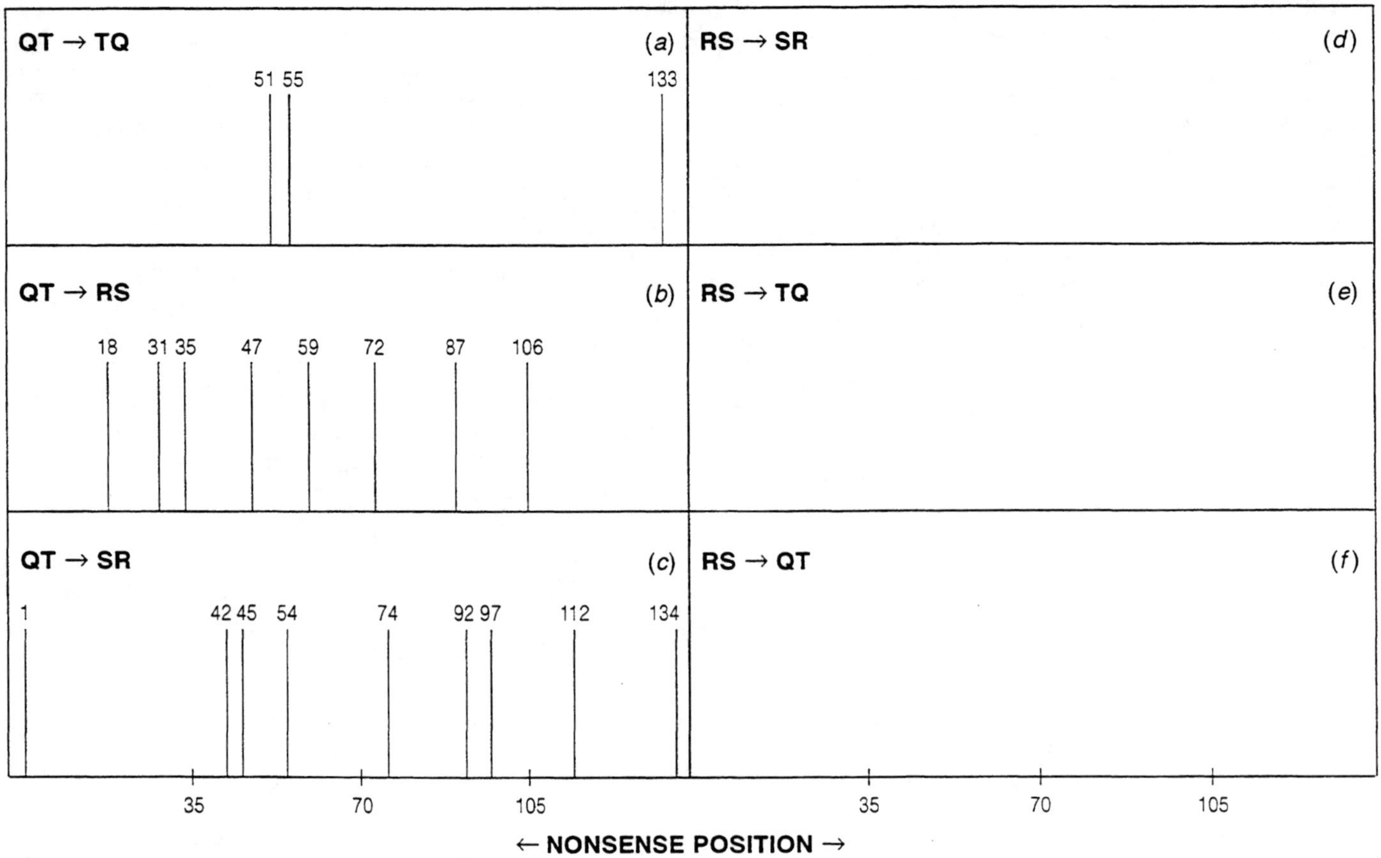

Figure 2

However, we really need to scan the other strand of the DNA for the same sequence configuration. Since we are not given this strand, we can instead look on the strand we have for the complement of the QRT sequence, which is QST. Since the T of the QRT sequence is methylated and mutates frequently, the Q that pairs with this T will also mutate frequently in the resulting double-stranded progeny. Thus, we must also look for QST sequences that can mutate to the nonsense codon **QSR** by changing the Q. Only QSQ codons in which the last Q is part of a QST sequence will be hotspots, so we are looking for a QSQ ST- sequence. Since the only QSQ codon (at position **106**) is indeed followed by an ST, then this codon will also mutate to nonsense at a higher rate and be a spontaneous hotspot.

In summary, there are expected spontaneous hotspots among the sites that go to nonsense at coding positions **55** and **106**. Figure 3 shows the approximate expected distribution.

2. (*a*) QSR reverts to the following sense codons:

SSR, TSR, RSR, QQR, QTR, QRR, QSQ, QST, QSS.

The base changes are as follows:

Q•T→S•R = SSR
Q•T→R•S = RSR
Q•T→T•Q = TSR
S•R→Q•T = QQR, QST
S•R→T•Q = QTR, QSQ
S•R→R•S = QRR, QSS

(*b*) To detect mutators in *Centauris polaris*, we need to be able to recognize cells that have a higher spontaneous mutation rate than normal. We can start with a strain carrying a nonsense mutation in the *K* gene. After mutagenesis, we plate on medium containing two different energy sources, crushed Voltor rat liver and comet dust, to increase the number of mutators. Colonies will grow up to a certain size and make green colonies utilizing the crushed Voltor rat liver. After they exhaust

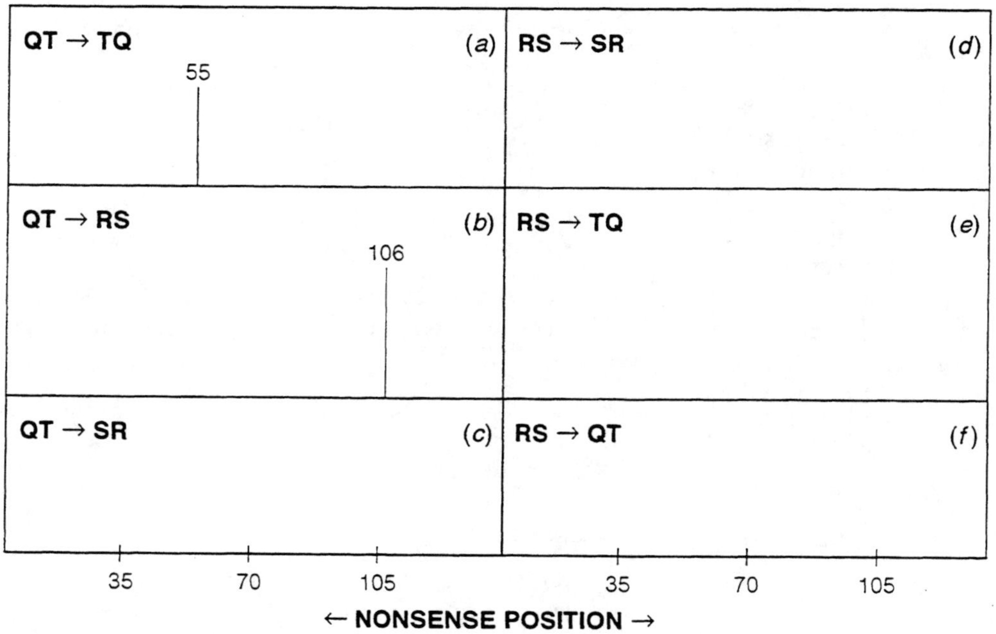

Figure 3

this energy source, they stop growing. Only the K^+ revertant cells in the colony can continue to grow, since only they can utilize the second energy source, comet dust. They will form yellow papillae growing out of the main green colonies. A mutator colony will have many more papillae than the other colonies.

3. In this experiment, we want to infect a cell carrying a deletion of the *K* gene with a phage that carries the Tn*FX* transposon with the *K* gene integrated in it, but we also want to have the infecting phage be composed of heteroduplex DNA in which one strand is from the wild-type *K* gene and the other has a point mutation in *K* (K^-). We would denature and anneal DNA from phage carrying either a K^+ or a K^- inserted gene. The K^+ and K^- phage would each carry one unique restriction site. After annealing, only the heteroduplex DNA would be resistant to cutting by both enzymes. Assume that the transposon also encodes an antibiotic. We would then infect cells and plate on minimal medium with asteroid dust and the antibiotic and look for sectored colonies that are part blue and part red. If transposition proceeds by a replicative mechanism, we would not expect to see sectored colonies, but if it occurred via a nonreplicative mechanism, then the heteroduplex DNA would be conserved and we would observe sectored colonies.

B. 1. (*a*) The following sense codons can go to the nonsense codon **DLG**:

GLG, MLG, LLG, DMG, DDG, DGG, DLD, DLL, DLM.

Each of the six base changes results in the following sense codons changing to the nonsense codon DLG, with the number indicating the codon, the first codon in the sequence being numbered "1."

D•G→G•D = GLG (51, 86), DLD (1, 3, 44, 59, 70)
M•L→D•G = MLG (43, 101), DLL (29, 31, 54, 72, 74, 90, 110)
M•L→G•D = LLG (38, 57, 64, 111, 113), DLM (24, 26, 47, 49, 100, 103)
M•L→L•M = DMG (67, 92, 114, 119)
D•G→M•L = DGG (9, 14, 22)
D•G→L•M = DDG (82, 106)

These can be arranged in a figure like Figure 1 in part A, which diagrams the position of each site.

(*b*) Since the mutagen makes all possible base changes at M•L base pairs, then only the sites represented by the DLM, LLG, DLL, MLG, and

DMG codons will be hit. These can be arranged in a figure like Figure 2 in part A.

(*c*) The expected results for spontaneous base substitutions would have each of the possible sites at approximately similar frequencies except for hotspots at methylated bases. These would occur at the M of a MDD sequence (read 5′→3′) or the L of a GGL sequence (see answer to 1c in part A). The only codon that could mutate to the **DLG** nonsense codon via a change at a methylated base is the DLM codon at position **49**, which would therefore show up as a hotspot.

2. (*a*) DLG reverts to the following sense codons:

   LLG, MLG, GLG, DMG, DGG, DDG, DLD, DLM, DLL.

   The base changes are as follows:

   D•G→L•M = LLG, DLM
   D•G→M•L = MLG, DLL
   D•G→G•D = GLG, DLD
   M•L→L•M = DMG
   M•L→D•G = DGG
   M•L→G•D = DDG

   (*b*) To detect mutator strains, we take any of the R^- mutant strains and mutagenize and plate on minimal medium with both comet dust and asteroid dust. Cells will form orange colonies, but only revertants can grow on the asteroid dust. They will form black papillae, and the colonies with very large numbers of papillae are mutators. These are purified and tested against each of the six mutations in R that were built in the last section.

3. The answer is identical to that for question 3 in part A except that the heteroduplex is R^-/R^+ and the medium used for the transduction is minimal medium with the antibiotic and Paramador rat liver. The presence of sectored orange and black colonies indicates conservative transposition.

C. 1. (*a*) The following sense codons can go to the nonsense codon **NJF**:

   XJF, FJF, JJF, NNF, NFF, NXF, NJJ, NJN, NJX.

   Each of the six base changes results in the following sense codons changing to the nonsense codon NJF, with the number indicating the codon, the first codon in the sequence being numbered "1."

   X•F→N•J = XJF (10, 12, 19, 46, 49, 63, 67, 120, 124, 134), NFF (41, 45)
   X•F→J•N = FJF (32, 35, 39, 54, 94, 106, 117), NXF (3, 17, 64, 84, 130)
   X•F→F•X = NJX (4, 65, 85, 91, 98, 128)
   N•J→J•N = JJF (101, 109, 144), NNF (13, 75)
   N•J→X•F = NJJ (2, 26, 81, 108)
   N•J→F•X = NJN (none)

   These can be arranged in a figure like Figure 1 in part A, which diagrams the position of each site.

   (*b*) Since the mutagen makes all possible base changes at X•F base pairs, then only the sites represented by the XJF, NFF, FJF, NXF, and NJX codons will be hit. These can be arranged in a figure like Figure 2 in part A.

   (*c*) The expected results for spontaneous base substitutions would have each of the possible sites at similar frequencies except for hotspots at the methylated bases. Here, we would scan for JNX sequences and the complement FJN to see whether mutating the J of the JNX sequence or the N of the FJN sequence would result in a nonsense **NJF** codon. To do this, we would look for NJJ codons that are part of an NJJ NX- sequence and we would note that the FJN sequence cannot yield an NJF codon by changing the N of the FJN sequence. The only NJJ NX- sequence is at position 2/3, so the NJJ codon at position **2** would be a spontaneous hotspot.

2. (*a*) NJF reverts to the following sense codons:

   XJF, JJF, FJF, NNF, NXF, NFF, NJX, NJJ, NJN.

   The base changes are as follows:

   X•F→N•J = XJF, NFF
   X•F→J•N = FJF, NXF
   X•F→F•X = NJX
   N•J→X•F = NJJ
   N•J→F•X = NJN
   N•J→J•N = JJF, NNF

   (*b*) We would mutagenize any of the strains from *a* above and plate on minimal medium containing both Paramador bat extract and Traxor soil mix. All colonies will be silver colored, but only revertants to Q^+ can grow further on the Traxor soil mix, making gold colored papillae arising out of the silver colored colonies. Those with more papillae than normal are mutators. These colonies would be picked and purified and tested against the other strains containing different mutations in the Q gene to determine the specificity of the mutator.

3. The answer is identical to that for question 3 in part A except that the heteroduplex is Q^+/Q^- and the medium is comet dust and the antibiotic. We would look for sectored colonies that are part white and part purple.

D. 1. (*a*) The following sense codons can go to the nonsense codon **PLC**:

GLC, CLC, LLC, PGC, PCC, PPC, PLG, PLL, PLP.

Each of the six base changes results in the following sense codons changing to the nonsense codon PLC, with the number indicating the codon, the first codon in the sequence being numbered "1."

G•C→L•P = CLC (85, 110, 120), PGC (2, 40)
G•C→P•L = GLC (10, 29, 30, 31, 46, 58, 62, 94), PCC (77)
G•C→C•G = PLG (9, 16, 26, 41, 48, 83, 87, 99, 101, 122, 130)
L•P→G•C = PLP (34, 89, 95, 124)
L•P→C•G = PLL (12, 21, 82, 133)
L•P→P•L = LLC (70), PPC (63, 78, 97, 101, 108)

These can be arranged in a figure like Figure 1 in part A, which diagrams the position of each site.

(*b*) Since the mutagen makes all possible base changes at L•P base pairs, then only the sites represented by the PLP, PLL, LLC, and PPC codons will be hit. These can be arranged in a figure like Figure 2 in part A.

(*c*) We need to scan the sequence for both PCL sequences, where the P will mutate frequently, and for the complement, the PGL sequence, where the L will mutate frequently. For the PCL sequence, only PLP CL- sequences will mutate frequently to create the **PLC** codon, and for the PGL sequence, only -PG LLC sequences will mutate frequently to PLC. We note that at position **34/35** we have the sequence PLP CLP, which is a hotspot for the change to PLC CLP. There are no other sequences that would give a spontaneous hotspot for the change to the PLC codon.

2. (*a*) PLC reverts to the following sense codons:

GLC, CLC, LLC, PGC, PCC, PPC, PLG, PLL, PLP.

The base changes are as follows:

G•C→L•P = PLP
G•C→P•L = PLL
G•C→C•G = PLG
P•L→G•C = GLC, PCC
P•L→C•G = CLC, PGC
P•L→L•P = LLC, PPC

(*b*) To detect mutators, we start out with a strain carrying a nonsense mutation in the T gene. After mutagenesis, we plate on medium containing two different energy sources, silicon dioxide and Traxor soil mix, to increase the number of mutators. Colonies will grow to a certain size and appear turquoise. After they exhaust this energy source, they stop growing. Only the T^+ cells in the colony can continue growing, since only they can use the silicon dioxide. They will form white papillae growing out of the main colonies. A mutator colony will have many more papillae than the other colonies.

3. The answer is identical to that for question 3 in part A except that the heteroduplex is T^+/T^- and the medium contains the antibiotic plus Paramador bat extract. The appearance of blue and red sectored colonies would indicate a conservative mode of transposition.

E. 1. (*a*) The following sense codons can go to the nonsense codon **KKM**:

LKM, JKM, MKM, KLM, KJM, KMM, KKK, KKJ, KKL.

Each of the six base changes results in the following sense codons changing to the nonsense codon KKM, with the number indicating the codon, the first codon in the sequence being numbered "1."

J•K→L•M = KKK (74, 103)
J•K→M•L = KKJ (31, 43, 91, 93)
J•K→K•J = JKM (7), KJM (10, 28, 52, 88, 119, 124)
L•M→J•K = MKM (16, 35, 69, 86, 122, 129), KMM (21, 24, 59)
L•M→K•J = LKM (17), KLM (1, 48)
L•M→M•L = KKL (13, 39, 66, 77, 116)

These can be arranged in a figure like Figure 1 in part A, which diagrams the position of each site.

(*b*) Since the mutagen makes J•K→L•M and J•K→M•L changes, only the sites represented by the KKK and the KKJ codons will be hit. These can be arranged in a figure like Figure 2 in part A.

(*c*) The J mutates frequently when part of the JLK sequence. Only the KKJ LK- sequence can result in a nonsense mutation **KKM** by a change at this J. If we scan the four KKJ sequences (that are in the correct reading frame), then we find that only the KKJ at position **91** is followed by a LK sequence. Thus, this codon will be a spontaneous hotspot. We must also scan for the complement to JLK, which is JMK, where the K mutates frequently. However, this sequence cannot be involved in generating the KKM sequence from a wild-type codon.

2. (*a*) KLM reverts to the following sense codons:

MLM, LLM, JLM, KJM, KLK, KKM, KLJ, KMM, KLL.

The base changes are as follows:

J•K→L•M = MLM
J•K→M•L = LLM
J•K→K•J = JLM
L•M→J•K = KJM, KLK
L•M→K•J = KKM, KLJ
L•M→M•L = KMM, KLL

(*b*) To detect mutators, we start out with a strain carrying a nonsense mutation in the *J* gene, mutagenize it, and plate the survivors on medium containing both meteor dust and Traxor ring worm extract. After exhausting the meteor dust, the colonies, which are black, will stop growing. Only J^+ revertants can continue to grow, since they can utilize the Traxor ring worm extract. They will form orange papillae. Mutator colonies will have more papillae than normal. These can be picked, purified, and retested for increased reversion to J^+.

3. The answer is identical to that for question 3 in part A except that the heteroduplex is J^+/J^- and the medium contains the antibiotic plus Traxor soil mix. The presence of green and yellow sectored colonies indicates a conservative mode of transposition.

F. 1. (*a*) The following sense codons can go to the nonsense codon **VVD**:

DVD, KVD, PVD, VPD, VKD, VDD, VVV, VVP, VVK.

Each of the six base changes results in the following sense codons changing to the nonsense codon VVD, with the number indicating the codon, the first codon in the sequence being numbered "1."

K•P→V•D = KVD (28, 65, 114), VKD (6, 30, 37, 76, 98, 112), VVP (43, 87)
D•V→V•D = VDD (1, 51), DVD (22), VVV (35, 85, 102)
K•P→D•V = VVK (4, 59, 70, 81, 109, 121), VPD (18, 55, 68, 79, 97, 100), PVD (53, 77, 119)

These can be arranged in a figure like Figure 1 in part A, which diagrams the position of each site.

(*b*) Since the mutagen makes all possible base changes at D•V base pairs, then only the sites represented by the VDD, DVD, and VVV codons will be hit. These can be arranged in a figure like Figure 2 in part A.

(*c*) When the K of a KVP sequence is methylated, then only VVK sequences can be hotspots if the VVK is followed by a VP-. This is true only for the VVK at position **81**, which will thus be a hotspot. However, we also need to look for the KDP sequence, where the P will mutate frequently. Any PVD sequences that are part of a -KD PVD sequence will be hotspots. This is true for the PVD sequence at position **77**.

2. (*a*) VVD reverts to the following sense codons:

DVD, KVD, PVD, VPD, VKD, VDD, VVV, VVP, VVK.

The base changes are as follows:

D•V→V•D = DVD, VDD, VVV
D•V→K•P = PVD, VPD, VVK
D•V→P•K = KVD, VKD, VVP

(*b*) To detect mutators, we start with a strain carrying a nonsense mutation in the *V* gene. We mutagenize the strain and plate on medium containing space ice and Voltor insect extract. Colonies will grow on the space ice and make yellow colonies until they exhaust the medium. V^+ revertants will form brown papillae by growing on the Voltor insect extract. Colonies with more papillae than normal are mutators. These are picked and analyzed.

3. The answer is identical to that for question 3 in part A except that the heteroduplex is V^+/V^- and the medium contains the antibiotic plus silicon dioxide. The presence of turquoise and gold sectored colonies indicates a conservative mode of transposition.

Scenario 13 / In Search of . . . Inhibitors

A. First, we have to be able to make Phy$^-$ derivatives of some of these strains. This can be done by selecting for W4 resistance, since some of the resistant mutants result from large deletions of the W4-*rec* gene that extend into and past the *phy* gene. We may also be able to detect deletions that cut out both the *phy* gene and the *ile* gene, which we can then employ in selections.

Second, we need to put the *phy* genes onto the F factor. The first step is to integrate the F factor into the W4-*rec* locus by selecting for W4-resistant cells in an F$^+$ strain and testing for those that are Hfr strains with an origin at that point by crossing with the multiply marked recipient and looking for strains that donate either *ile* early and *met* late or vice versa. We can then take either of these Hfr strains and mate them with a Rec$^-$ derivative of the multiply marked strain that we have made Phy$^-$. Note that this strain is already Ile$^-$. We would then select for donation of Ile$^+$, since F′ factors derived from this Hfr that carry *ile* must also carry *phy*. We do not need the Rec$^-$ strain if we use the Hfr that donates *phy* and *ile* late and do a short, 30- to 60-minute, mating and interrupt and then look for Ile$^+$ cells. Since the Hfr cannot donate these markers in either of the above crosses, inheritance of both Ile$^+$ and Phy$^+$ must be due to an F′ *ile-phy*, which can be verified by transferring this to a recipient in a short mating. Instead of selecting for Ile$^+$, we can look for Phy$^+$ directly by screening colonies on grape extract infusion medium for those that are red, although this would be much more laborious, since the desired colonies would be rare.

The next step is to take the episome we have constructed and put it into the deletion strain we have made. Here, we would use either a *phy* deletion derivative of the multiply marked recipient, which is already Ile$^-$, or a deletion we isolated that deletes both *phy* and *ile*. We introduce the episome and plate at high temperature (preventing replication of the F′ factor outside the chromosome) on plates seeded with W3 phage. We can either use grape extract infusion medium and look for red colonies, or if we have made an F′ *ile-phy*, we can plate on minimal medium without isoleucine and then test colonies on grape infusion medium. The single event that will produce red colonies at 42°C in the presence of W3 phage is an integration of the F factor into the W3-*rec* locus, which would then place the *phy* and also the *ile* gene near the L20 attachment site.

The last step is to lysogenize with phage L20 and then UV-induce and take the lysate and transduce the Ile$^-$ Phy$^-$ strain to Ile$^+$ and screen the transductants for those that are Phy$^+$. Or, we can transduce a Phy$^-$ strain and plate on grape infusion medium and look for red colonies, although this would be laborious since these would be rare. The Phy$^+$ colonies are then reinduced and the new lysate is again used to transduce to demonstrate high-frequency transduction of the *phy* gene.

B. The first step would be to isolate mutants that result in the Phy$^-$ phenotype. Mutagenized cultures would be plated on grape extract infusion medium, and white colonies would be picked and purified for further study. Mapping experiments would establish that, as expected, one set of mutations mapped in the *phy* locus, the gene that encodes the PHY protein. However, another group of mutants would map in a different gene, which we would call the *R* gene, the positive regulator of the *phy* gene. Inactivating the positive regulator would result in the failure to synthesize PHY even in the presence of the inducing tanins. Complementation experiments would show that R^+ is dominant to R^-, since heterodiploids that were R^+/R^- would now synthesize PHY.

In analogy with experiments done with the *Escherichia coli ara* operon, as well as with the CAP factor for the *lac* operon, we would look for mutants that become Phy$^+$ in the presence of an R^- allele. In other words, we would start with an R$^-$ strain that has a mutation in the gene encoding the positive regulator and look for revertants that can now make red colonies on grape infusion medium. In fact, to make it easier to detect mutations in the control region, we could even begin with a partial diploid that is R^-/R^- and look for revertants that make white colonies on grape infusion medium. These revertants would have *cis*-dominant mutations in the control region, similar to the *araI*^{c} mutations (or the *UV5* mutation in *lac*), that can now synthesize some level of PHY even in the absence of a positive regulatory protein. We might also call these mutations in the initiator part of the control region I^c mutations. In heterodiploids of the configuration I^c phy^+/I^+ phy^-, the cells would synthesize PHY in response to inducing tanins, but in heterodiploids that were I^c phy^-/I^+ phy^+, the cells would not synthesize PHY even in response to tanins, reflecting the *cis*-dominant nature of the mutation in the control region.

C. 1. Introducing a charged residue such as arginine into the interior of a protein can have disastrous effects on the stability of the protein, since polar groups are excluded from the interior, which almost always consists of hydrophobic residues.

2. Although changes on the exterior of a protein are usually tolerated, some residues are involved in subunit contacts, and these residues may not be easily substitutable. Here, the serine might make a hydrogen bond in a subunit contact that the glycine cannot make.

3. Even in some surface positions, such as here, proline can disrupt α helices and result in a defective protein.

4. Binding sites for ligands, or catalytic centers of enzymes, are often in the crevice. The arginine might be part of the active site or the binding site, and changing it to lysine may eliminate the very specific contacts involving the arginine.

5. Although some changes from one nonpolar residue to another are tolerated in the interior, sometimes the contacts there are very specific. Here, the isoleucine might be making contacts that the smaller valine cannot make, and the loss of these bonds might destabilize the protein.

Scenario 14 / Mission Impossible: Attack of the Oil-eating Clones

A. We should note that a misprint in the strain list on page 653 of the text lists Slm⁻ in place of Grs⁻. Let's consider Slm⁻ equivalent to Grs⁻ in this problem. The first step here is to isolated an F′ that carries the *grsE* gene. To do this, we need to integrate F near the *grsE* gene and then detect rare mutants that have excised F incorrectly, bringing the *grsE* genes with them. We note that the gene encoding the receptor site for the K7 phage is near the *grsE* gene, so we select for integrations of F into the K7-*rec* gene in strain 1 by growing on medium with acridine orange on plates seeded with K7 phage. We can test for whether we have an Hfr strain with its origin of replication at the K7-*rec* locus by mating with strain 2 and looking for early donation of Thy⁺ and late donation of His⁺ or vice versa. We then can use strain 3 as a recipient in a mating with either Hfr and select for Grs⁺ on medium with crude oil as the only carbon source. Since the chromosomal region for *grsE* cannot be incorporated into the recipient chromosome in a Rec⁻ strain, only the rare F′ factors carrying the *grsE* gene can confer the Grs⁺ phenoptype.

The second step involves putting the F′ constructed above into strain 4 and then selecting for colonies that can grow on medium with crude oil as the only carbon source in the presence of acridine orange on plates seeded with J19 phage. The only single event that can result in colonies will be an integration of the F′ *grsE* into the J19-*rec* locus, which positions the *grsE* gene near the attachment site for phage ϕ39. The next step is to lysogenize with ϕ39 and then to UV-induce and use the lysate to transduce strain 4, selecting for Grs⁺ colonies on medium with crude oil as the only carbon source. Colonies that grow are induced again and the lysate again used to transform strain 4, which it should do at high efficiency.

B. As a first step, we would isolate mutants that are GrsE⁻. We can apply a penicillin selection, as discussed in Unit 4 (after the analysis of the papers from the Yanofsky group) to enrich for Grs⁻ cells after mutagenesis. We could simply replicate mutagenized cells after mutagenesis onto medium with and without crude oil, but that would be very laborious, and a penicillin enrichment is necessary. In addition to the *grsE* gene, the gene encoding greasase, we would also find an unlinked gene, which we can call *R*, that when inactivated prevented the expression of the *grsE* gene, since the activator made by this gene is a required positive control factor. Complementation tests would show that *R*⁺ is dominant to *R*⁻. We would also select for revertants from R⁻ strains that were now R⁺. Some of these should contain mutations in the initiator region that are *cis*-dominant, since they do not need the activator for the genes linked to this altered initiator region (see the answer to Scenario 13B).

We could detect evidence for negative control by doing a version of the Pajamo experiment. We would use a recipient deleted for the *grsE* gene and for a presumed linked negative control gene, *I*. Note that the activator gene, *R*, is unlinked. When we mate an *I*⁺*grsE*⁺ Hfr with this recipient and follow the production of greasase, we would see that in the absence of inducer, there is uncontrolled synthesis of the enzyme, which is then shut off after one or two hours, although it can be induced with inducers to the full rate at any point. Since the activator is present in the F⁻, it will allow synthesis of the enzyme in the absence of the *I* product, which in this case is an enzyme that destroys the inducer. This argues that there is a gene product encoded by *I* that must build up in the cytoplasm in order to block greasase synthesis. One can look for mutants that are constitutive. With the tools that are given, this would be laborious, since it would involve taking mutagenized cells and testing individual colonies for high levels of the enzyme in the absence of the inducer. The constitutive mutants, or I⁻, would behave just as those in the repressor model, namely, *I*⁺ would be dominant to *I*⁻ in diploid tests.

C.

1. Although nonpolar residues can often be exchanged for certain other nonpolar residues, some substitutions are not tolerated because the contacts in the interior are specific for the wild-type amino acid. Here, the smaller alanine would probably not be able to make all of the contacts that the leucine could.

2. Most changes on the exterior of the protein are tolerated, but some residues on the exterior may be involved in subunit interactions, which may be the case here, and the substitution could lead to loss of subunit aggregation.

3. The same explanation is valid here as for number 2 above, namely, a subunit contact may in-

volve the serine residue, which is lost when tyrosine is substituted for the serine.

4. Proline can cause disruptions of the secondary structure by interrupting α helices. This would cause a defective protein in many cases, such as here.

5. The crevice of the protein usually contains the active site of an enzyme or the binding site for a ligand for other proteins. Residues that either contact the ligand, participate in the reaction, or maintain the architecture of the binding site are critical and are usually sensitive to substitutions. One possibility is that the isoleucine is making interior contacts that form the framework of the crevice binding site, and substituting even a valine might alter the framework.

Scenario 15 / Mission Impossible: The Hunt for Red *mouR*

A. To construct an F′ factor carrying the *mouR* gene, we first have to isolate a strain with the F factor integrated near the *mouR* gene. We can take the wild-type strain carrying the temperature-sensitive F factor and select at high temperature for strains that are resistant to phage K9, since resistant mutants will have mutations or interruptions in the K9-*rec* gene, which is near *mouR*. Although single-step mutations can be resistant, some will result from the integration of the F factor into the K9 gene. We can find this fraction by testing resistant mutants for those that are Hfr strains with their point of origin at K9-*rec* by using the multiply marked Str^r recipient that is supplied. The correct Hfr will donate either Leu^+ or Thr^+ efficiently, but not the other marker.

To isolate an F′ *mouR*, we could take one of the Hfr strains identified in the previous step and cross it with strain 3, which is Rec^- and $MouR^-$ and look for rare colonies that are white on wine medium. These colonies could only result from the transfer to the Rec^- strain of an F′ *mouR*. However, these colonies would be very rare. A better way is to take the Hfr that donates Leu^+ early and Thr^+, and thus $MouR^+$, late and to cross it in a short (30-minute) mating to the multiply marked strain 2, selecting for Thr^+. Since the *thr-mouR* region is not brought in by the Hfr in a short mating, only an F′ can transfer this region in a 30-minute mating. The F′ *thr* must also carry *mouR*, which can be verified by testing whether the Thr^+ colonies turn white on wine medium.

The next step is similar to that from Beckwith and Signer. We need to select for strains in which the F′ *mouR* has integrated into the N5-*rec* locus, since it is near the attachment site for the temperate phage φ77. If we are using the F′ *mouR* that is temperature-sensitive, then we would use strain 4, which is deleted for the *mouR* gene, and select colonies that are phage φ77-resistant at high temperature and are white on wine medium. (If we use the F′ *thr mouR*, then we need to use strain 3, which is Rec^-, to prevent the F′ from reintegrating into the *thr* locus by homologous recombination.) The surviving colonies should have the F′ *mouR* integrated into the N5-*rec* locus, which can be verified by showing that they are Hfr strains with their integration point at the N5-*rec* locus. The multiply marked recipient strain (#2) is used in a cross with the prospective Hfrs, and the number of His^+ and Thy^+ colonies are scored. The correct Hfr should give high His^+ and low Thy^+ or vice versa.

The next step is to lysogenize with the temperate phage φ77 and then to induce the phage, presumably with UV, and use the resulting lysate to transduce strain 4 to $MouR^+$. Here, we cannot directly select, so we would have to screen colonies for those that are white on wine medium. These would be rare, but they could then be purified, and cultures could be induced again with UV, this time yielding a high frequency of $MouR^+$ colonies.

B. The first step would be to isolate mutants that result in the $MouR^-$ phenotype. Mutagenized cultures would be plated on wine medium, and red colonies would be picked and purified for further study. Mapping experiments would establish that, as expected, one set of mutants mapped in the *mouR* locus, the gene that encodes the enzyme moutonase. However, another group of mutants would map in a different gene, which we will call the *R* gene, the positive regulator of the *mouR* gene. Inactivating the positive regulator would result in a failure to synthesize moutonase even in the presence of the inducer Mouton Rothschild. Complementation experiments would show that R^+ is dominant to R^-, since heterodiploids that were R^+/R^- would now synthesize mountonase.

In analogy with experiments done with the *Escherichia coli ara* operon, as well as with the CAP factor for the *lac* operon, we would look for mutants that become $MouR^+$ in the presence of an R^- allele. In other words, we would start with an R^- strain that has a mutation in the gene encoding the positive regulator and look for revertants that can now make white colonies on wine medium. In fact, to make it easier to detect mutations in the control region, we could even begin with a partial diploid that is R^-/R^- and look for revertants that make white colonies on wine medium. These revertants would have *cis*-dominant mutations in the control region, similar to the $araI^c$ mutations (or the *UV5* mutation in *lac*), that can now synthesize some level of moutonase even in the absence of a positive regulatory protein. We might also call these mutations in the initiator part of the control region I^c mutations. In heterodiploids of the configuration I^c $mouR^+/I^+$ $mouR^-$, the cells would synthesize moutonase in response to Mouton Rothschild, but in heterodiploids that were I^c *mouR*$^-$/I^+ *mouR*$^+$, the cells would not synthesize moutonase even in response to Mouton Rothschild, reflecting the *cis*-dominant nature of the mutation in the control region.

C. 1. Residues in the crevice are often involved in the active site. The charged residue glutamic acid changing to a similarly charged residue aspartic acid could cause a defective protein if the glutamic acid was involved in the active site or the architecture of the active site.

2. Changes on the exterior of the protein are often tolerated, but proline can disrupt the secondary structure by breaking up an α helix and can result in a destabilized or defective protein.

3. Changing a glutamine to tyrosine on the exterior could cause a defective protein if the glutamine was part of a subunit-subunit interaction.

4. Changing a nonpolar residue to a polar residue in the interior of the protein will usually destabilize the protein, since polar residues are excluded from the hydrophobic interior.

5. Although certain changes of one nonpolar residue to another nonpolar residue are often tolerated in the interior of a protein, this is not always the case. Here, the isoleucine may be making a contact that the leucine residue cannot make, thus destabilizing the structure.

Scenario 16 / The Swiss Connection: Trace the Mutants

A. The first step is to isolate a strain carrying an F′ with the *neuR* gene. To do this, we first need to integrate the F factor near the *neuR* gene, creating an Hfr strain. We can do this by selecting for P9-resistant cells in the presence of acridine orange in strain 1, since some of the resistant colonies will result from the integration of F into the P9-*rec* locus. We can find these by screening for Hfr strains that donate either Leu$^+$ early and Arg$^+$ late or vice versa by crossing the P9-resistant colonies with strain 2. We then can mate either of these Hfr strains with strain 3, which is Rec$^-$ and thus cannot incorporate genes from an Hfr into its chromosome. The only way to inherit the NeuR$^+$ character would be to receive an F′ carrying the *neuR* genes. We would plate the mating on indicator plates with streptomycin to kill the donors and look for white colonies. Because we do not have a selection, these colonies would be rare. Alternatively, we might use strain 2 and one of the Hfr strains that donates Leu$^+$ as a late marker and select for Leu$^+$ in a short, 30- to 60-minute, interrupted mating. Here, we would be looking for an F′ that has the *leu* genes and thus would also carry the *neuR* genes. The F′ would be large, but we would have a selection for it.

The second step would be to use strain 4 and select for B46-resistant colonies in the presence of acridine orange, which are white on indicator medium. Some of these will result from the integration of the F′ into the B46-*rec* locus. These colonies can be identified as Hfr strains donating either His$^+$ early and Met$^+$ late or vice versa. The next step is to infect with phage ϕ17, since its attachment site is near the B46-*rec* locus, and then to induce with UV and use the lysate to transduce strain 4, screening for white colonies on indicator medium. These colonies will be very rare. Alternatively, if one uses the larger episome carrying the *leu* genes, then one could select for transducing phage carrying the *leu* genes by using strain 2 and then screening for white colonies among these on indicator medum. If the distance between the two genes is too great, these phage may not be found.

B. The answer to this problem is identical to that for Scenario 14B except that here the inducer is the half-toxin and the NeuR$^-$ colonies turn red and those that are NeuR$^+$ turn white on the indicator medium used here. The NeuR$^-$ colonies can be found by screening mutagenized cultures for those that form red colonies on indicator medium in the presence of the inducer (half-toxin). Constitutive colonies are those that are white on indicator medium even in the absence of inducer.

C.

1. Although substitutions on the exterior of a protein are usually tolerated, some sites on the exterior make contacts with other subunits. Here, the serine to glycine change might affect subunit association.
2. Glycine cannot make the contacts that alanine and other nonpolar residues can make and will leave a space in the interior. This can destabilize the structure.
3. The introduction of a charged residue such as histidine into the interior of the protein can cause destabilization.
4. The crevice of the protein contains the active site or binding site for a ligand. Some residues are involved directly in the binding or indirectly by maintaining the architecture of the binding site. The phenylalanine might be involved in maintaining this framework and might make contacts that valine cannot make.
5. Proline can disrupt α helices and cause a distortion of the secondary structure. The introduction of proline leads to a defective protein in many positions, including those on the exterior of the protein.

Scenario 17 / Outbreak

The virus generates inserts by transposing a segment of its genome into chromosomal genes. The facts given suggest that transposition occurs via an RNA intermediate. Transcription generates a primary transcript from which the 2000-bp intron is spliced. The processed RNA is then reverse-transcribed to generate a double-stranded DNA copy that is then inserted into the genome of the host. We could prove this mechanism by placing a known intron in the middle of the inserted segment and then observing whether it appears in the inserted DNA that we find transposed into the chromosome. In analogy with the experiments of Boeke et al., we might use the selection for His^+ revertants in a strain containing the promoter deletion in the *HIS3* gene, since most of the revertants contain insertions upstream of this region. We would then examine the transposed DNA in the *HIS3* promoter region.

Appendix: Student Solutions

The following pages present actual student solutions to a few scenarios as examples of the level of detail that can be expected when these problems are given as take-home exams.

Scenario 1 / Chances Are It's Poisson

H. Since there are four \$100,000 slots, we will divide the wheel up into twenty four-slot intervals. That is, all the \$100,000 slots now represent a single interval in our distribution. The rest of the slots account for the other 19 intervals, which are all the same as far as we are concerned. Since we are dividing the 1120 balls (one per spin) between the twenty intervals we find:

$$m = \frac{\text{No. Balls}}{\text{Intervals}} = \frac{1120}{20} = 56$$

To find the Poisson distribution we apply the following relationship:

$$P_n = \frac{m^n e^{-m}}{n!}$$

where m is the average number events per interval calculated above, e is the base of the natural logarithm, and P_n is the probability that and interval will have n occurrences of the event in question.

For example the probability P_0 that and interval will have zero events in 1120 spins is

$$P_0 = e^{-m} = e^{-56} = 4.7808 \times 10^{-25}$$

similarly

$$P_1 = \frac{m^n e^{-m}}{n!} = \frac{56e^{-56}}{1} = 2.6773 \times 10^{-23}$$

the probability that the \$100,000 prize would come up only twenty times would be:

$$P_{20} = \frac{m^n e^{-m}}{n!} = \frac{56^{20} e^{-56}}{20!} = 1.8078 \times 10^{-8}$$

the results Poisson distribution is calculated from this scenario for n = 0 to 20 follow

n	P_n
0	$4.78 \text{ X } 10^{-25}$
1	$2.68 \text{ X } 10^{-23}$
2	$7.5 \text{ X } 10^{-22}$
3	$1.4 \text{ X } 10^{-20}$
4	$1.96 \text{ X } 10^{-19}$
5	$2.19 \text{ X } 10^{-18}$
6	$2.05 \text{ X } 10^{-17}$
7	$1.64 \text{ X } 10^{-16}$
8	$1.15 \text{ X } 10^{-15}$
9	$7.14 \text{ X } 10^{-15}$

10	$4 .00\text{X } 10^{-14}$
11	$2.03 \text{ X } 10^{-13}$
12	$9.49 \text{ X } 10^{-13}$
13	$4.09 \text{ X } 10^{-12}$
14	$1.64 \text{ X } 10^{-11}$
15	$6.11 \text{ X } 10^{-11}$
16	$2.14 \text{ X } 10^{-10}$
17	$7.04 \text{ X } 10^{-10}$
18	$2.19 \text{ X } 10^{-09}$
19	$6.46 \text{ X } 10^{-09}$
20	$1.81 \text{ X } 10^{-08}$

The sum of these probabilities equals 2.77 X 10^{-8} and represents the total probability that $100,000 will come up twenty times or less. Again, the probability that it would come up exactly twenty times is 1.81 X 10^{-8}. Either of these is **extremely** unlikely, and this essentially would not happen if the spins were truly random. In other words, the big spin people are probably taking some nice all-expenses-paid vacation with the money they were supposed to be awarding!!!

We follow the same process for the million dollar prize. Since there are only two million dollar slots we divide the wheel into forty two-slot intervals. Therefore

$$m = \frac{1120}{40} = 28$$

For n = 0 to 14

n	P_n
0	6.91 X 10^{-13}
1	1.94 X 10^{-11}
2	2.71 X 10^{-10}
3	2.53 X 10^{-09}
4	1.77 X 10^{-08}
5	9.92 X 10^{-08}
6	4.63 X 10^{-07}
7	1.85 X 10^{-06}
8	6.48 X 10^{-06}
9	2.02 X 10^{-05}
10	5.64 X 10^{-05}
11	0.000144
12	0.000335
13	0.000722
14	0.001444
sum	0.00273

The probability that the million dollar prize will come up exactly fourteen times is about 0.14%. The probability that the one million dollars will come up fourteen times or less is about 0.27%. These results are orders of magnitude more likely than the results of the $100,000 prize. However, they are still very improbable. Conclusion: somebody had better check the Big Spin Wheel or arrest some state officials.

Scenario 9 / Mission Impossible: To Decode Deadly Genotypes

B. Since our agent possessed only the strains Arg+ and Arg-, we can easily see that the code will be a binary one similar to that used to encode computer data. That is each letter of our message will be represented by what would be analogous to a byte of computer information, which would itself be comprised of a string bits. Each of these bits is either a plus or a minus. We determine the length of each byte by calculating the minimum length required to encode for all twenty-six letters of the alphabet. We find this length to be five, which gives us a total of thirty-two possible "words." (Four would be two few giving us only sixteen) The general rule that can be used to determine the number of possible combinations is simply given by the number of "letters" possible raised to length of each "word." In our case it is 2^5 or thirty-two.

We then divide the string of pluses and minuses into distinct groups of five, being sure to not to omit or add any in order to avoid the very troublesome frameshift errors. A number is assigned to each possible combination of pluses and minuses simply because they are more familiar and easier to deal with (a table showing these assignments is attached). We now have a string of numbers, each representing a letter of the alphabet and a piece of our message:

```
15  21  17  12  24  21  17  08  22  22  21  14
18  17  21  22  21  24  08  16  21  08  14  23
21  10  12  08  04  14  21  03  18  18  07  18
17  21  17  07  19  20  21  18  17  12  18  23
21  10  07  14  06  20  18  07  12  01  08  10
05  10  21  06  21  14  22  08  18  08  20  08
14  18  17  21  23  12  16  13  20  21  12  18
07  01  17  21  04  04  07  10  10  20  04  07
27  16  21  22  08  12  10  18  17  21  16  07
22  21  21  08  06  17  18  21  21  14  04  08
06  17  18  20  08  03  18  21  21  14  10  21
01  18  18  15  21  14  18  27  07  14  21  04
08  06  17  18  18  07  22  08  20  12  04  05
```

We then tabulate the frequency of each "letter" and "double letters":

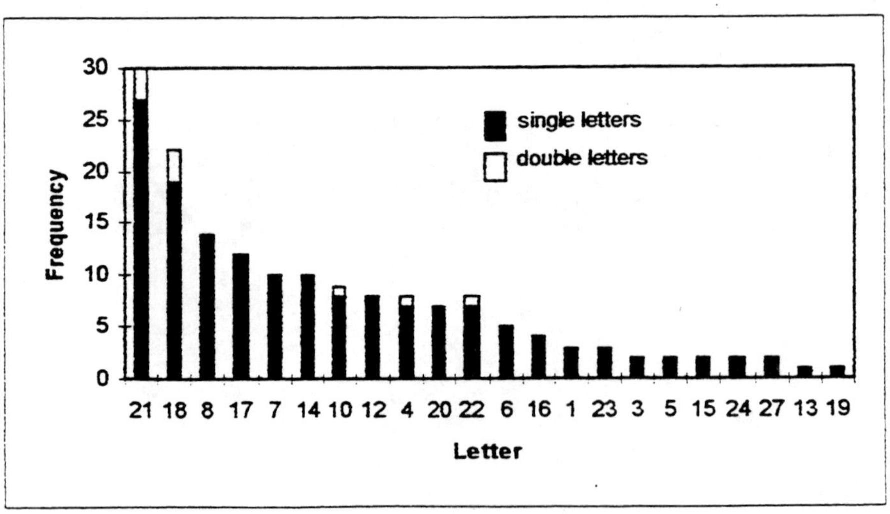

We can use the standard frequencies for letters (Valerio, Romanini, Thomas) as a guide to give us a clue as to what the letters may be based solely upon their frequency. However, we must bear in mind that our message is only 156 letters long, which is a relatively small sample size in comparison to the millions of letters which the standard frequencies represent. Therefore, we cannot expect the letter frequencies to match precisely. Additionally, there are certain pitfalls of which we need to be aware when dealing with double letters, bigrams, and two letter words, for example. Since the words are in a continuous line, we cannot initially be sure if a particular pair of letters is at the interior of a word or the last and first letter of two adjacent words (assuming there are no spaces in our message). This is also true for longer strings of letters: THE could either be simply the word itself, a word ending in TH followed by one starting with E, or a word ending in T followed by one starting with HE. However, the most likely candidates for potential double letters would be SS, EE and TT. As we shall see later two of these letter pairs turn up in our message.

In any case, there seems to a general consensus that E and T are the two most common letters we start by replacing them for 21 and 18. Additionally, there are four occurrences of the pattern 18 17 21 or T_E. Knowing that THE is the most common three letter word, we can replace H for 17. With just three letters we have already cracked 37% of the message. However, none of the choices we have made up until this point has completely airtight, and in fact, the process is mainly one of trial and error, guided by the frequency information given and a bit of common sense about the combinations possible in the English language (nothing new for any avid crossword puzzle zealot). The fruits of our labor would yield the following:

W	E	H	A	V	E	H	I	D	D	E	N
T	H	E	D	E	V	I	C	E	I	N	B
E	L	A	I	R	N	E	X	T	T	O	T
H	E	H	O	U	S	E	T	H	A	T	B
E	L	O	N	G	S	T	O	A	F	I	L
M	L	E	G	E	N	D	I	T	I	S	I
N	T	H	E	B	A	C	K	S	E	A	T
O	F	H	E	R	R	O	L	L	S	R	O
Y	C	E	D	I	A	L	T	H	E	C	O
D	E	E	I	G	H	T	E	E	N	R	I
G	H	T	S	I	X	T	E	E	N	L	E
F	T	T	W	E	N	T	Y	O	N	E	R
I	G	H	T	T	O	D	I	S	A	R	M

Translated: "We have hidden the device in Bel Air next to the house that belongs to the house that belongs to a film legend. It is in the back seat of her Rolls Royce. Dial the code eighteen right, sixteen left, twenty-one right to disarm." In which case, we had better hurry up to save our alma mater from certain destruction!!!

The Cracked CODE

#	Code	Frequency	Letter
1	+++++	3	F
2	++++−	0	
3	+++−+	2	X
4	++−++	7	R
5	+−+++	2	M
6	−++++	5	G
7	+++−−	10	O
8	++−+−	14	I
9	+−++−	0	
10	−+++−	8	L
11	++−−+	0	
12	+−+−+	8	A
13	−++−+	1	K
14	+−−++	10	N
15	−+−++	2	W
16	−−+++	4	C
17	++−−−	12	H
18	+−+−−	19	T
19	−++−−	1	U
20	+−−+−	7	S
21	−+−+−	27	E
22	−−++−	7	D
23	+−−−+	3	B
24	−+−−+	2	V
25	−−+−+	0	
26	−−−++	0	
27	+−−−−	2	Y
28	−+−−−	0	
29	−−+−−	0	
30	−−−+−	0	
31	−−−−+	0	
32	−−−−−	0	

Scenario 10 / Jurassic Engineering

B. 1. Having already saved Bel Air, UCLA, and a good portion of the Los Angeles metropolitan area from nuclear devastation, stopping Karpak and his reptiles is a less daunting task because the missions are in some ways similar. This time our "alphabet" represents the thirty-four possible amino acids (twenty normal and fourteen dinosaur). Instead of pluses and minuses for our "code letters," we have the six possible nucleotide bases: A, T, C, G, X, Y (we must bear in mind the difference between letters of the "alphabet" and those of the code). We use the same formula to determine the minimum word length. With six bases a two-nucleotide word length should be sufficient to code for our thirty-four amino acids since 6^2=36. The additionally leaves two codons free for nonsense stop codons.

We use the data from hydroxylamine mutagenized Tyrannosaur 4 virus to determine which amino acid codons contain Gs and Cs and whether the Cs are on the sense or anti-sense strand of the DNA. The number of different amino acids that are produced by a mutagenized codon in metanogen B is equal to the number of G-C pairs in the original codon. One is the result of a mutated C on the sense strand. The other is the result of a mutagenized C on the anti-sense strand. Since all mutants still grow on methanogen B, we should find all possible amino acid substitutions on that strain. However, since only mutations on the antisense strand will pass through strain K, we know that the original sequence has a C in the antisense strand, and therefore a C in the codon. If no mutant species are found in strain K, then we know that the C is on the sense strand and the codon contains a G. From this we find that that following codons (the mRNA sequence complementary to the sense strand) contain Gs or Cs (not in order):

PSR	C
VEL	CG
HPR	G
ARG	G

SCY	G
TYR	CC
TRP	CG
NOR	GG

CYS	G
CGL	G
PTR	C
PRO	C

GLY	G
HYP	G
SMT	C
BRO	C

We can also gather information from the amino acid substitutions that have resulted from the amino acid substitutions. If they occurred from a mutation in the sense strand (in B but not K), it means that the mRNA is transcribed directly from the U' base in the sense strand and that the codon contains an A. If the mutation occurred on the anti-sense strand, the complementary sense DNA would contain an A and the mRNA codon would contain a U. From this information we find that the following contain As and Us:

PHE	U
HYP	UG
SMT	AC
LEU*	AA
VAL*	AU

ILE	U
ALA	U
PTR	AC
ARG	UG
SCY	AG

HPR	AG
PTY	A
GLN	A
MET	U
SER	A

ASN*	AU
GLU	U

The codons of the amino acids with an asterisk to the right were found to contain both A and U (depending on the original codon, they fit into both categories described previously).

In the case of LEU, there two As because it was produced by both a mutagenized HPR codon and a mutagenized SCY codon and in each case only appeared in strain B. Bases listed in italics were deduced in previous steps. We also know that these mutant codons share one base pair in common with the wild-codon since only one base is changed in any given case.

Since 5-azacytosine causes the a G-C to C-G base substitution in both strands it can only tell use that the mutagenized and the original codon each contains a C or a G. Again we also know that the mutant codons share one base in common with the wild type codon since only one base pair is changed in each case. Much of what we find is duplicate information, so the only additional information we have learned is that the following contain a C or a G: ILE, ALA. If we look at the data, we find that mutation to ARG by hydroxylamine gives VAL while 5-azacytosine gives ILE. The three must share one base in common, and we see that ARG contains UG while VAL contains AU. Therefore, the common base is U and ILE also contains a C. In a similar manner we find that ALA contains a U and a C.

Finally, we know that a mutagen that changed A to X resulted in the substitution of LEU for SER in on of the proteins. It doesn't tell us whether it was on the sense or anti-sense strand, but since we have already determined that the codon for LEU is AA, we know that the mutation occurred on the antisense strand (the wild-type sense strand would be TT). We can now deduce that the SER codon contains an X since the anti-sense strand contains X and the replicated sense strand would contain Y. MET, GLY and PRO must also contain X because we can deduce from the mutagenisis results that they all share one base in common.

From all the accumulated information we have:

PSR	C*
VEL	CG
HPR	AG
ARG	UG
SCY	AG
TYR	**CC**
TRP	CG
NOR	**GG**

CYS	G*
CGL	G+
PTR	AC
PRO	CX
GLY	GX
HYP	UG
SMT	AC
BRO	C+

PHE	U*
LEU	**AA**
VAL	AU
ILE	UC
ALA	UC
PTY	A*
GLN	A+
MET	UX

SER	AX
ASN	UA
GLU	U+

Note that the tables give the base composition of each codon but not the order of the bases. We are sure of the order and the compositions of codons in boldface since they are composed of two of the same base. Symbols represent unknown bases that certain codons must have in common.

The results of the frameshift mutation can elucidate the rest of the puzzle:

WILD TYPE PROTEIN:	LEU	ASP	PSR	GLN	VEL	HPS	ARG	LYS	VAL	GLU	PSR	PTY
	AA	UU	YC	XA	CG	GA	GU	AY	AU	XU	YC	YA
	AU	UY	CX	AC	GG	AG	UA	YA	UX	UY	CY	AG
FRAMESHIFT PROTEIN:	VAL	PTE	PRO	PTR	NOR	SCY	ASN	PTY	MET	PTE	HIS	SCY

HPR	SCY	SER	LEU	HPR	TYR	TRP	NOR	CYS	CGL	ARG	SER	PHE	...
GA	AG	AX	AA	GA	CC	GC	GG	YG	XG	GU	AX	YU	
AA	GA	XA	AG	AC	CG	CG	GY	CX	GG	UA	XY		
LEU	HPR	GLN	SCY	PTR	VEL	VEL	APA	GLY	NOR	ASN	STOP		

We can immediately fill in AA, CC and GG for LEU, TYR, and NOR respectively since we are sure that they are appropriate. Since we know that the mutation was a single deletion frameshift at the first base we can begin filling in bases for the codons adjacent to the ones whose bases we already know. In conjunction with the information we have found previously we can begin to deduce the order as well as the composition of the codons. For example, from the first set of codons in the diagram above, we know that the first letter of VAL must be the second A of LEU. We already know from the tables above that VAL contains both A and U, so the codon must be AU. From the tables above, we know that ASN also contains A and U, so the it's codon must be UA since AU is already taken by VAL. We now also know that the first letter of ASP must be U. We continue with the same logic, filling the rest of the diagram. Our compiled results:

Jurassic Genetic Code

Second Letter

First Letter

	U	C	A	G	X	Y
U	ASP	ALA	ASN	HYP	MET	PTE
C	ILE	TYR	SMT	VEL	PRO	HIS
A	VAL	PTR	LEU	SCY	SER	LYS
G	ARG	TRP	HPR	NOR	GLY	APA
X	GLU	BRO	GLN	CGL		STOP
Y	PHE	PSR	PTY	CYS		

2. There are still three possible codons for which we have not found a corresponding amino acid. We have no found the codon for THR and JUR. The remaining codon could be another nonsense codon or could be a second codon for an amino acid. We could find the remaining codons by continuing the mutigenization experiments with the mutagen that alters A to X.. We can observe the changes in the protein sequence as we had done previously. Missing codons can be found as follows:

Mutagenized PTE and PHE will result in the YY codon:

```
          PTE
 anti: TY      YY
sense: AX --> XX
 mRNA: UY      YY

          PHE
 anti: YT      YY
sense: XA --> XX
 mRNA: YU      YY
```

Mutagenized SER and GLN will result in the XX codon:

```
          PTE
 anti: AX --> XX
sense: TY      YY
 mRNA: AX      XX

          PHE
 anti: XA --> XX
sense: YT      YY
 mRNA: XA      XX
```

Mutagenized VAL and MET will result in the YX codon:

```
          VAL
 anti: YA --> YX
sense: XT      XY
 mRNA: AU      YX

          MET
 anti: TX      YX
sense: AY --> XY
 mRNA: UX      YX
```

We can complete our Jurassic Genetic Code by simply analyzing the substitutions (or premature terminations) that result from our procedure. Karpak's nefarious endeavor is thwarted!!!

Scenario 12 / Hitchhiker's Guide to the Galaxy: Follow That Sequence

B1a. The following figure shows which one-step mutations could lead to the nonsense codon DLG:

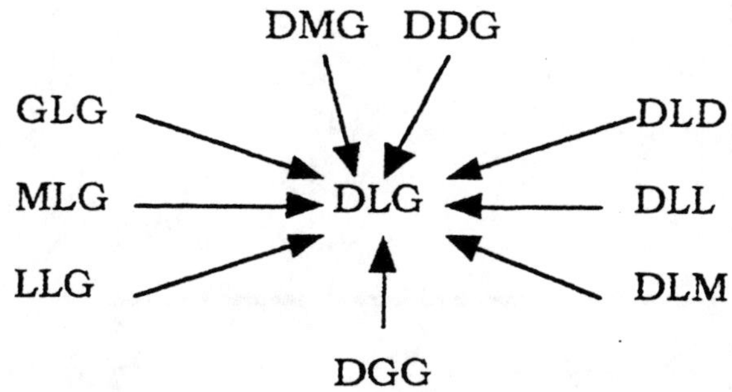

Using this we can take the sequence and search for all of the sites with one of these codons. They are organized according to type of change.

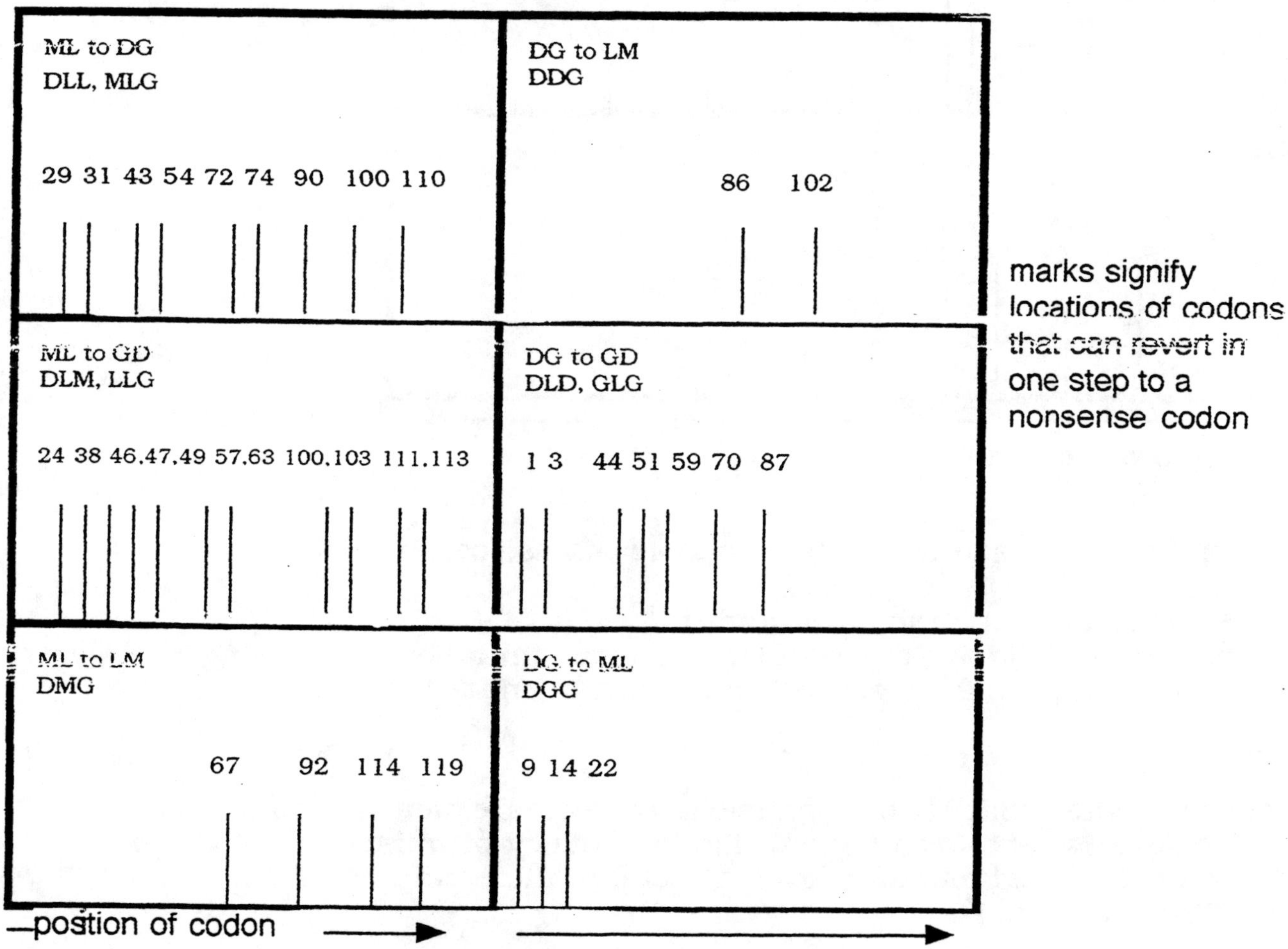

1b. You treat your cell with the mutagen that substitutes ML base pairs with any other type. The data below is what you would expect.

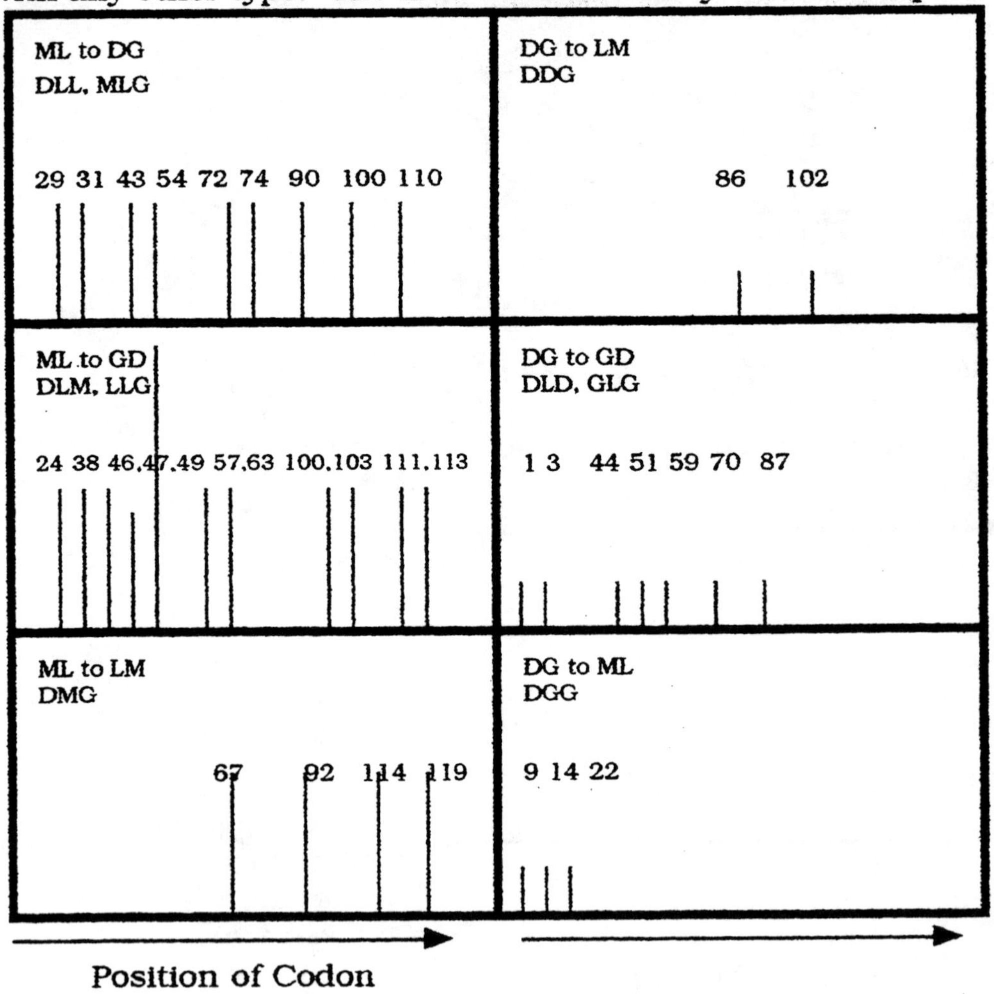

Marks represent the frequency of mutation to a nonsense codon

Because MDD is a mutation hotspot, and thus so is GGL, the triplet causes codon 49 to be a particularly common mutation None of the GGL triplets intersect with the potential mutation sites.

The system will show that ML base pairs change more frequently than others. A comparison between the mutation frequency with the mutagen, with the basal mutation level is useful in understanding the potency of the mutagen.

1c. The spontaneous mutation rate would resemble the chart in 1a except the Y axis would represent frequency. The distribution of nonsense mutations over the possible frequencies should follow a Poisson distribution. Below it is shown as if all of these codons are bascially of the same frequency when it comes to the change to a nonsense mutation.

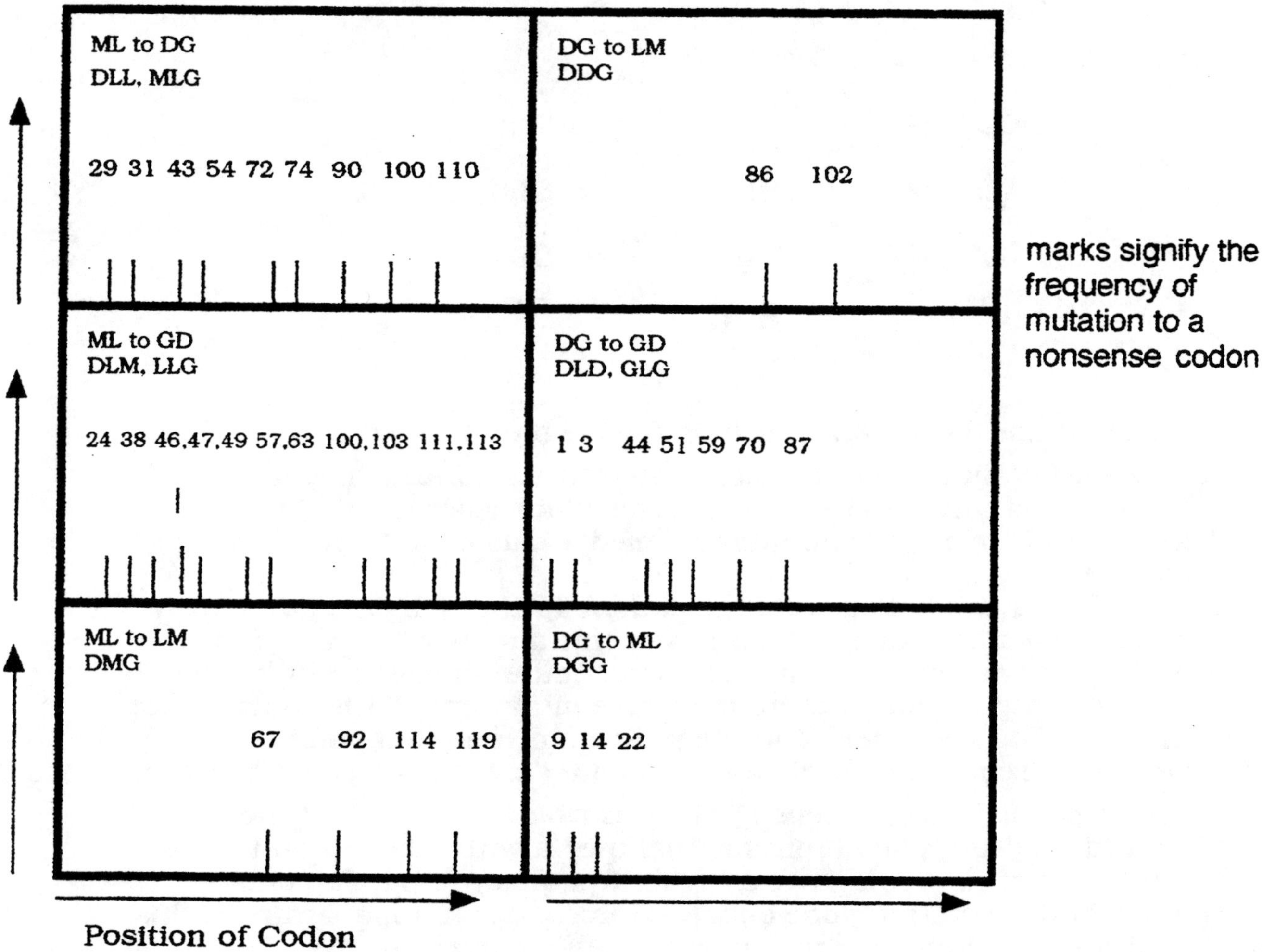

2a. The following chart is similar to that in 1a. GD to LM changes result in DLL and MLG codons. GD to ML changes result in the sense codons DLM and LLG. The ML to LM transversion could make a DMG, the ML to GD a DDG, the ML to DG a DGG, and the DG to GD tranversion could result in a DLD or GLG. This changes are summarized below.

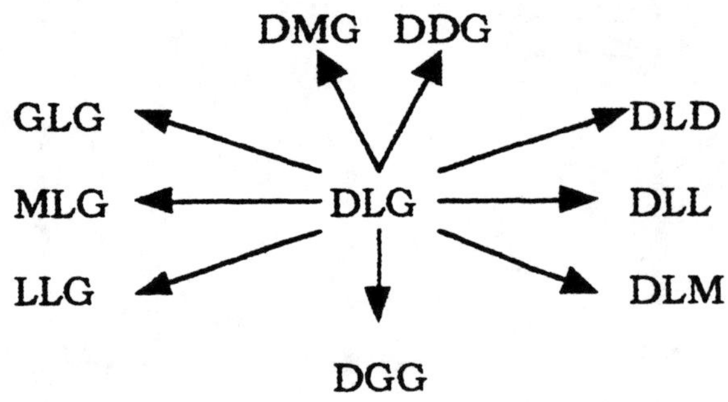

2b. How do you use the Traxor *C. polaris* R gene mutant system to detect mutators, knowing that R^- mutants cannot grow on minimal media with asteroid dust, upon which the wild type forms black colonies, and knowing that the R- mutants can grow on minimal media plus comet dust forming orange colonies?

One would do a papillation test as was done in the Nghiem et al. paper. If in the problem when it says that "you have r mutants of different types" it means that you have more than just nonsense mutations, you would begin by characterizing those strains that are nonsense mutations. This could be done by sequencing. If on the other hand the problem means that you have many types of R nonsense mutations, then the step just mentioned could be skipped. The nonsense codon in this case is DLG. You would take each of these R^- strains individually and plate it on minimal media with asteroid and comet dusts. Mutator strains would have a higher frequency of papillation. Papillation occurs when, after one energy source is used up, a strain, in this case a revertant, continues to grow using another energy source. Growth on the different energy sources in this system yields different colored plaques. Simply for this experiment, you would choose those strains that consistently show the most black colonies growing on the larger orange ones and those would be your mutator strains. These mutator strains would then be studied for defects in repair pathways

3. Trying to determine whether the TnFX transposes in a replicative or non-replicative manner.

1. You take the phages carrying the R+ and R- TnFX transposons and get DNA from them.
2. The DNA from these two phage types then would be mixed and annealed to form a heteroduplex, where one strand was the DNA from the R+ and the other from the R-.
3. Infect this in one of the R mutant strains used in question 2. Plate this on Paramador rat liver.
4. If you do not get sectored colonies, then the mechanism is replicative. Unsectored colonies tell you that all of the cells got only one type of gene present in the heteroduplex. Therefore the DNA replicates itself in the transpostition process. It inserts indirectly. Sectored colonies tell you that in the insertion of the transposon no information is lost. When the cell divides, one cell gets only the R- genes. The other gets only the R+ genes. If you got unsectored colonies of two types then you would know that the mechanism is semiconservative, in that the recipient cell gets one or the other gene in the heteroduplex.

The diagram below recaplitulates the chief aspects of the description above.

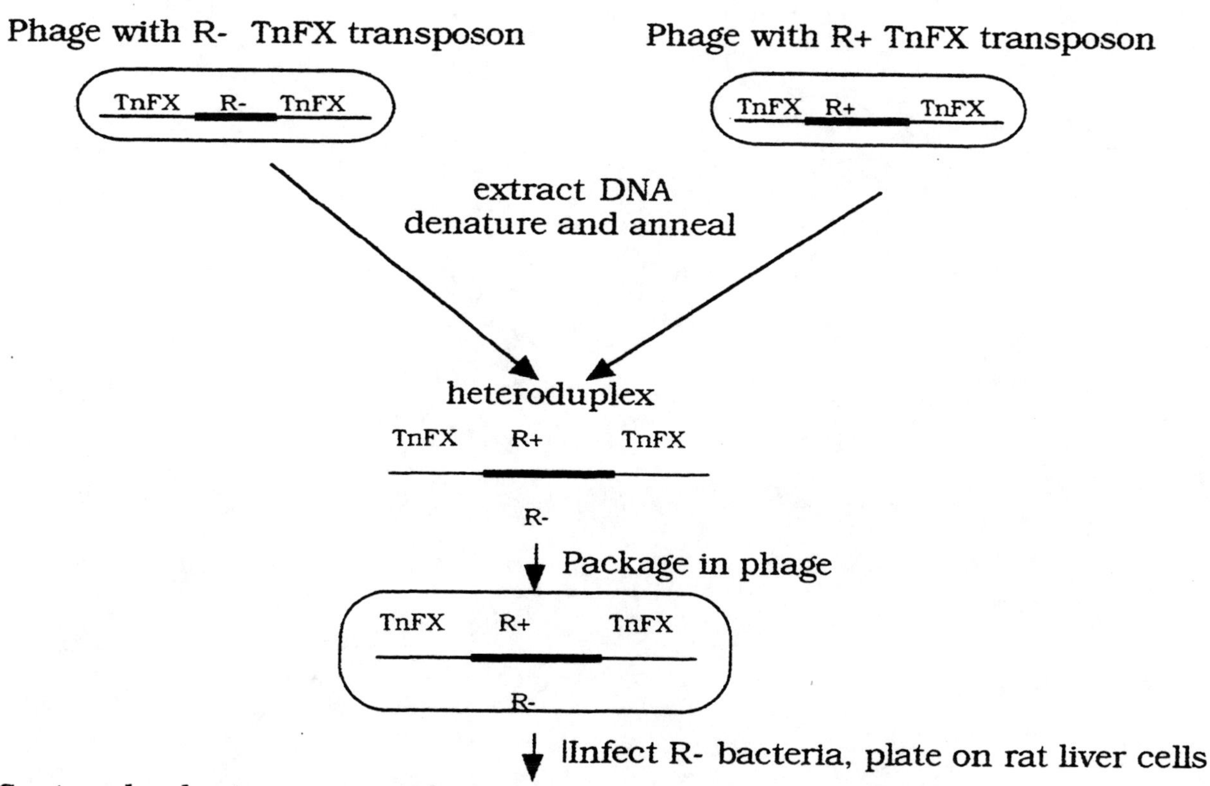

Sectored colonies= non replicative, unsectored colonies=replicative